FRONTIERS IN ELECTRONICS

Advanced Modeling of
Nanoscale Electron Devices

SELECTED TOPICS IN ELECTRONICS AND SYSTEMS

Editor-in-Chief: **M. S. Shur**

---

*The complete list of the published volumes in the series can be found at
http://www.worldscientific.com/series/stes

Selected Topics in Electronics and Systems – Vol. 54

FRONTIERS IN ELECTRONICS

Advanced Modeling of Nanoscale Electron Devices

Editors

Benjamin Iñiguez
Universitat Rovira I Virgili, Spain

Tor A Fjeldly
Norwegian University of Science and Technology (NTNU), Norway

World Scientific

NEW JERSEY · LONDON · SINGAPORE · BEIJING · SHANGHAI · HONG KONG · TAIPEI · CHENNAI

Published by

World Scientific Publishing Co. Pte. Ltd.

5 Toh Tuck Link, Singapore 596224

USA office: 27 Warren Street, Suite 401-402, Hackensack, NJ 07601

UK office: 57 Shelton Street, Covent Garden, London WC2H 9HE

British Library Cataloguing-in-Publication Data
A catalogue record for this book is available from the British Library.

Selected Topics in Electronics and Systems — Vol. 54
FRONTIERS IN ELECTRONICS
Advanced Modeling of Nanoscale Electron Devices

Copyright © 2014 by World Scientific Publishing Co. Pte. Ltd.

All rights reserved. This book, or parts thereof, may not be reproduced in any form or by any means, electronic or mechanical, including photocopying, recording or any information storage and retrieval system now known or to be invented, without written permission from the publisher.

For photocopying of material in this volume, please pay a copying fee through the Copyright Clearance Center, Inc., 222 Rosewood Drive, Danvers, MA 01923, USA. In this case permission to photocopy is not required from the publisher.

ISBN 978-981-4583-18-3

Printed in Singapore

PREFACE

Accurate modeling of nanoscale electron devices is essential for both their technological improvement and optimization, and for their utilization in circuit design. Depending on the application, different levels of modeling have to be used. Technology oriented design requires highly physical models for both transport and electrostatics. Circuit design needs analytical compact models derived by considering a number of approximations.

This book on Advanced Modeling of Nanoscale Electron Devices, consists of four chapters to address at different modeling levels for different nanoscale MOS structures (Single- and Multi-Gate MOSFETs). The collection of these chapters attempts to provide a comprehensive coverage on the different levels of electrostatics and transport modeling for these devices, and the relationships between them. In particular, the issues of quantum transport approaches, analytical predictive 2D/3D modeling and design-oriented compact modeling are considered. It should be of interest to researchers working on modeling at any level, providing them with a clear explanation of the approaches used and the links with modeling techniques for either higher or lower levels.

List and short description of the chapters:

F. Gámiz, C. Sampedro, L. Donetti, A. Godoy, "Monte-Carlo Simulation of Ultra-Thin Film Silicon-on-Insulator MOSFETs"

This chapter reviews the basis of the Multi-Subband Monte Carlo (MSB-MC) method. In the first part of the chapter, the authors present a comprehensive coverage of the impact of different Buried OXide (BOX) configurations on the scaling of extremely thin fully depleted SOI devices using a Multi-Subband Ensemble Monte Carlo simulator (MS-EMC). In the second part of the chapter, the authors use the MS-EMC simulator to present a study of the electron transport in ultrashort DGSOI devices with different confinement and transport directions.

R. Clerc, G. Ghibaudo, "Analytical Models and Electrical Characterisation of Advanced MOSFETs in the Quasi-Ballistic Regime"

This chapter reviews the current understanding of the quasi-ballistic transport in advanced MOSFETs, underpinning the derivation and limits of corresponding analytical models. Furthermore, in order to estimate the "degree of ballisticity" achieved in advanced technologies, the chapter presents the different strategies used to compare these models and experiments.

T. A. Fjeldly, U. Monga, "Physics-Based Analytical Modeling of Nanoscale Multigate MOSFETs"

This chapter describes the derivation of a comprehensive modeling framework for double-gate and gate-all-around MOSFETs based on a conformal mapping analysis of the subthreshold potential distribution in the device body, and a self-consistent procedure in the above threshold regime. In an alternative modeling framework, covering a wide range of multigate devices in a unified manner, the potential distribution is derived from a select set of isomorphic trial functions that reflect the geometry and symmetry properties of the devices.

B. Iñiguez, R. Ritzenthaler, F. Lime, "Compact Modeling of Double and Triple Gate MOSFETs"

This chapter presents new techniques for the compact modeling of Double- and Triple-Gate MOSFETs. First of all, full analytical compact expressions for the charges and the drain current for each of the two channels of symmetrical Double Gate MOSFETs UTB SOI and Asymmetric Double Gate MOSFETs with independent gate operation are discussed. In the second part, the authors develop a framework for a fully 3D compact modeling of the electrostatics of Tri-Gate MOSFETs, including the back gate bias and short-channel effects. Finally, it further demonstrates that without the use of any fitting parameter, the model can be extended to nearly all the MuGFETs devices (GAA/ΠFETs/TGFETs/FinFETs/symDGFETs/FDSOI planar devices), and therefore could potentially be used as a core model for the scaling and calibration of a wide range of MuGFETs.

EDITORS
Benjamin Iñiguez
Tor A. Fjeldly

CONTENTS

MONTE-CARLO SIMULATION OF ULTRA-THIN FILM SILICON-ON-INSULATOR MOSFETs

FRANCISCO GÁMIZ, CARLOS SAMPEDRO, LUCA DONETTI, and ANDRES GODOY

Research Group on Nanoelectronics, University of Granada, Facultad de Ciencias, Avd.Fuentenueva, s/n
18071 Granada, Spain
fgamiz@ugr.es

State-of-the-Art devices are approaching to the performance limit of traditional MOSFET as the critical dimensions are shrunk. Ultrathin fully depleted Silicon-on-Insulator transistors and multi-gate devices based on SOI technology are the best candidates to become a standard solution to overcome the problems arising from such aggressive scaling. Moreover, the flexibility of SOI wafers and processes allows the use of different channel materials, substrate orientations and layer thicknesses to enhance the performance of CMOS circuits. From the point of view of simulation, these devices pose a significant challenge. Simulations tools have to include quantum effects in the whole structure to correctly describe the behavior of these devices. The Multi-Subband Monte Carlo (MSB-MC) approach constitutes today's most accurate method for the study of nanodevices with important applications to SOI devices. After reviewing the main basis of MSB-MC method, we have applied it to answer important questions which remain open regarding ultimate SOI devices. In the first part of the chapter we present a thorough study of the impact of different Buried OXide (BOX) configurations on the scaling of extremely thin fully depleted SOI devices using a Multi-Subband Ensemble Monte Carlo simulator (MS-EMC). Standard thick BOX, ultra thin BOX (UTBOX) and UTBOX with ground plane (UTBOX+GP) solutions have been considered in order to check their influence on short channel effects (SCEs). The simulations show that the main limiting factor for downscaling is the DIBL and the UTBOX+GP configuration is the only valid one to downscale SGSOI transistors beyond 20 nm channel length keeping the silicon slab thickness above the theoretical limit of 5 nm, where thickness variability and mobility reduction would play an important role. In the second part, we have used the multisubband Ensemble Monte Carlo simulator to study the electron transport in ultrashort DGSOI devices with different confinement and transport directions. Our simulation results show that transport effective mass, and subband redistribution are the main factors that affect drift and scattering processes and, therefore, the general performance of DGSOI devices when orientation is changed

Keywords: Silicon-on-Insulator; scaling; short-channel effects; Multisubband Ensemble Monte Carlo; multigate transistor; quantum effects.

1. Introduction

New fabrication techniques and device concepts have been developed in the last 45 years to overcome the technological challenges that appear in order to follow Moore's Law and to fit the requirements of the International Technology Roadmap for the Semiconductor Industry (ITRS)[1]. Bulk MOSFET is still the main actor for the 32 nm node; however, as the channel length is reduced even more, short channel effects (SCEs) and variability problems arising from a highly doped channel[2] will make very difficult to maintain the standard bulk technology. In the search for the ideal device to be addressed for the 22 nm node and beyond, different options are under study. One of the possibilities to overcome these issues is the use of multiple gate devices (MuGFETs) which are able to extend the end of the roadmap thanks to an outstanding SCEs control[3, 4]. Nevertheless, the decision of moving into 3D device architectures mass production is a real breakthrough and implies important efforts from an economical and a technological point of view. An intermediate solution, which has not been yet fully exploited, is to take advantage of the benefits that Silicon-On-Insulator (SOI) provides to planar technology. Among the main SOI features, we can highlight a better control of short channel effects, lower parasitic capacitance, and greater tolerance to radiation; moreover, all this is achieved while maintaining compatibility with existing silicon fabrication facilities. To the afore commented advantages of SOI technology, ultrathin Fully Depleted SOI (FDSOI) devices add a simpler fabrication process than in the standard technology, since some steps like HALO implants or recessed channel techniques are not necessary anymore. This fact allows a reduction in the overall cost and an almost straightforward layout transfer from bulk to SOI[5]. Single gate (SG) SOI structures are addressed for this solution since they can provide an extra control of SCEs, small V_{th} variability and a reduction in the leakage current. In SOI devices the channel is kept virtually undoped[6] minimizing the impact of random dopant fluctuations (RDF). In this way, the two main problems of bulk MOSFET technology for sub-32 nm technological nodes are overcome.

The real interest of the semiconductor community on SOI devices arises when the thickness of the silicon layer is reduced to the order of ten or less nanometers. From the simulation and modeling point of view, such ultra-thin SOI devices require the development of new scattering and transport models, because those used for bulk or for thick SOI devices where the two Si-SiO$_2$ interfaces can be de-coupled, must be modified. Novel effects, such as volume inversion[7] and sub-band modulation effect[8] appear, strongly modifying the scattering mechanisms, and therefore the carrier mobility. We have deeply studied during last years these effects in long channel devices (both n- and p-type channels). For extended reviews of the properties of electrons and holes in ultrathin SOI devices and see how they are affected by the silicon layer thickness and crystallographic orientation of the devices, the interested reader can check References 8-13 and references therein. In those works, we extensively showed that in ultra-thin SOI layers, carrier behavior is affected by the presence of the two potential barriers at the Si-SiO$_2$ interfaces. Electron wavelength in the confinement direction is comparable to the spatial confinement length: hence quantum effects are produced by both geometric and electric field confinement. Therefore, quantum effects are not only as important as in bulk inversion layers, but they also depend on silicon layer thickness and can give rise to novel effects such as volume inversion[7]. This phenomenon occurs when the distance between the potential wells at each interface is small and inversion carriers are not confined to one of the interfaces but are spread throughout the whole channel. This affects not only the total carrier density in the channel, but also, and notably, their distribution and, as a consequence, their scattering rates. Quantum effects are highly dependent on the confinement effective mass of carriers; therefore, they are strongly affected by substrate crystal orientation. As a consequence, volume inversion effects change for different orientations and, clearly, depend on the type of carrier, i.e. electrons or holes. Figure 1 (left) shows the evolution of hole mobility with the silicon thickness in a DGSOI device, with different crystallographic orientations[13].

Certainly, the thickness of the silicon slab affects carrier distribution and carrier scattering, but we have also shown that phonon distribution in the structures is also strongly affected by the silicon thickness. Basically, there are two principal phenomena that modify the process of carrier scattering with the lattice vibrations in very thin silicon layers. First, the reduction of the electron momentum dimensionality and the momentum and energy conservation laws produces a drastic increase in the phonon scattering rate as the silicon thickness decreases[8]. The second phenomenon arises due to the modifications of the phonon modes caused by the acoustic and dielectric mismatches between the silicon and the silicon dioxide. These changes in properties give rise to confined and interface phonons in quantum wells, quantum wires and quantum dots[14]. The first phenomenon, increase in the phonon scattering rate in ultrathin silicon-on-insulator inversion produces a strong reduction in the mobility[10].To assess the second phenomenon we have developed a model to describe the quantization of acoustic phonon modes due to spatial confinement in silicon nanolayers using different structures and different boundary conditions[15,16]. Phonon quantization is included and the dispersion relations for distinct phonon modes are computed. This allows us to obtain the confined phonon scattering rates and to compute, using Monte Carlo simulations, the carrier mobility in ultrathin silicon on insulator inversion layers. We have shown that it is important to take into account acoustic phonon confinement in ultrathin silicon-on-insulator inversion layers by comparing electron mobility in ultrathin DGSOI devices, calculated assuming the usual bulk acoustic phonon model and also using the confined acoustic phonon model. Significant mobility reductions are obtained when the confined acoustic phonon model is used, even when other scattering mechanisms such as surface roughness scattering are taken into account. It is therefore essential to include such a model in the electron transport simulations of ultrathin SOI devices if we want to reproduce the actual behavior of electrons in silicon layers of nanometric thickness. Figure 1-(right) shows electron mobilities calculated by Monte Carlo simulation as a function of the silicon thickness in a double-gate SOI transistor at room temperature, taking into account the confined-phonon model.

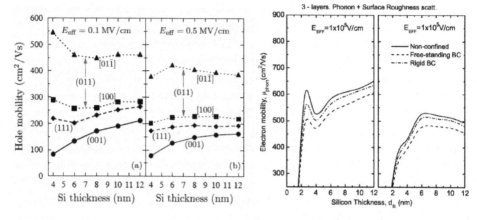

Fig. 1. (left) Effective hole mobility as a function of silicon thickness for different confinement and channel orientations (taken from Reference 13) (right) Monte-Carlo calculated electron mobility in double gate silicon-on-insulator inversion layers vs silicon thickness; surface roughness scattering has been taken into account in addition to phonon scattering (taken from Reference 15).

Up to now, we have considered the calculation of carrier transport properties in a boundless semiconductor structure, and therefore, we used a single particle Monte Carlo method. Thus we have showed the behavior of carrier transport in ultrathin SOI structures. However when we need to simulate the whole behavior of a realistic device, the single particle Monte Carlo approach is not appropriate because: i) the motion of particles is spatially restricted in the device region, so we need to set up suitable boundary conditions for the particles, and ii) the Boltzmann transport equation must be self-consistently solved with the Poisson and Schrödinger equations.

Different approaches from classical to full quantum can be considered depending on the needed accuracy, the computational resources and time available to perform the simulations. For the study of ultrashort channel devices, semi-classical approaches should not be used since confinements effects are of special importance[17]. At the opposite end of the spectrum, full quantum simulators based on numerical solutions of the Schrödinger equation or the Non-Equilibrium Green's Functions theory (NEGF) have also been developed[18]. In a quantum model, the transport of charged particles is treated coherently according to a quantum wave equation. In the simplest case of a single-particle Hamiltonian, carriers are considered as non-interacting waves described by the Schrödinger equation. The introduction of scattering in the simulations involves a very high computational cost and for this reason, only simplified models can be used in practical quantum simulations[18].

Between these extreme approaches, Ensemble Monte Carlo (EMC) simulators have been widely used since they present several advantages compared to full quantum approximations, such as a reduced computational cost, the possibility of considering a wide variety of scattering mechanisms and high accuracy. In order to include quantum effects in EMC codes, two main solutions are proposed in the literature. The first one consists of adding a quantum term to the electrostatic potential in order to correct it. The aim of this correction is to mimic the electron concentration profile obtained when the Schrödinger equation is solved. The calibration of such corrections generally needs a set of fitting parameters (e.g. carrier effective mass in the density gradient model) but the results obtained are still accurate from the transport point of view for devices with silicon thicknesses as low as a few nanometers[19]. The most commonly used approaches include the Density Gradient[20], the Effective Potential[21], or the Multi-Valley Effective Conduction Band-Edge method (MV-ECBE)[22]. The second solution is obtained from the coupling of the Monte Carlo solution of the Boltzmann Transport equation and the 1D solution of the Schrödinger equation in the confinement direction, evaluated in different slices of the device. This method, the so-called Multi Subband Monte Carlo (MSB) approach[23,24,25,26] provides what is to date the most detailed description of carrier transport in the device, since the scattering rates are obtained from a quantum solution. In this review, we have used the Multisubband Ensemble Monte Carlo method to study:

(1) the possibilities that the combined use of Ultrathin Buried Oxide (UTBOX) and a ground plane (GP) offers to determine whether the scaling of ultrathin single-gate FDSOI devices is useful to fulfill the requirements of sub-32 nm nodes reducing the impact of thickness fluctuation effects.
(2) the orientation effects in ultra-short channel DGSOI devices.

2. Ensemble Monte Carlo simulators

Ensemble Monte Carlo (EMC) simulators are widely used since they present several advantages compared to full quantum approximations. A reduced computational cost, the possibility of considering a wide variety of scattering mechanisms and high accuracy for devices with silicon thicknesses as low as a few nanometers[19] are some of the advantages of such simulators. As mentioned above to include quantum effects in EMC codes, two are the main solutions proposed in the literature:

(1) The addition of a correction term to the electrostatic potential, i.e. quantum correction, to mimic the carrier concentration profile obtained when the Schrödinger equation is solved in the structure. Different models have been developed following this philosophy, giving a good accuracy-computational cost ratio. The most commonly used approaches include the Density Gradient[20], the Effective Potential[21] or the Multi-Valley Effective Conduction Band-Edge method (MV-ECBE)[22].

(2) The coupling of the Boltzmann Transport Equation (BTE) solved by the Monte Carlo method in the transport plane with the Schrödinger equation in the confinement direction evaluated in different slices of the considered device. This method, called Multi Subband Ensemble Monte Carlo approach (MSB-EMC)[23,24,25,26] provides what is to date the most detailed description of carrier transport in the device, since the scattering rates are obtained from a quantum solution.

In the following sections, the different EMC approaches to the quantum problem will be presented and compared.

2.1. Quantum correction methods

The easiest way to include quantum effects in semi-classical simulators such as EMC codes is by adding a correction term to the electrostatic potential obtained from the solution of Poisson's equation. These approaches are widely used in MC codes since the early 2000s due to their computational efficiency (close to their semiclassical counterpart) and good accuracy. However, the practical implementation needs, in general, a previous calibration to obtain the set of fitting parameters to obtain accurate results from the transport point of view.

2.1.1. The effective potential method

The effective potential model[21] obtains the corrected potential from the convolution of the electrostatic potential, $V(x_0)$, and a Gaussian function which represents the "effective size" of the particle. The one dimensional form of the correction is given by:

$$V_{eff}(x) = \frac{1}{\sqrt{2\pi}a_0} \int V(x') \exp\left(-\frac{(x-x')^2}{2a_0^2}\right) dx'$$

(1)

where a_0 represents the spreading of the wave-packet. The maximum of the carrier distribution is then shifted from the oxide interface, reproducing the total inversion charge. However, the charge profile results incorrect when is compared to the 1D Schrödinger solution[20]. Therefore, the method is not appropriate when magnitudes that depend on overlapping integrals involving envelope functions extracted from carrier distributions need to be calculated. This is the case of surface roughness models or inter-valley scattering rates when size quantization is taken into account.

2.1.2. The density gradient method

A better approximation obtained from the Wigner potential approximation is the Density Gradient (DG)[27]. This method, which is one of the first quantum corrections implemented on a simulation code, produces very good results especially in drift diffusion simulations. The correction term is given by the following expression:

$$V_q = 2b_n \frac{\nabla^2\left(\sqrt{n}\right)}{\sqrt{n}}$$

(2)

where

$$b_n = \frac{\hbar^2}{4r\,m_n^*} \tag{3}$$

$\hbar$ is the reduced Planck's constant, m_n^* represents the electron effective mass and r is a parameter whose value varies from 1 for pure states (low temperatures or very strong confinement) to 3 for mixed states (high temperatures or weak confinement). The last two parameters are used to fit the carrier profile to the obtained from the solution of the 1D Schrödinger equation.

Figure 2 shows a comparison of the electron concentration profile obtained from the solution of the Schrödinger equation (black), the Density Gradient approach (red) and the effective potential (blue) for a MOS structure[20]. As can be observed, although both approximated techniques fit the inversion charge, in the case of the Density Gradient, it is also possible to reproduce the inversion charge profile.

The drawback of density gradient approach is the difficult implementation of the model in EMC codes. The main reason is the dependence of the driving force with the third derivative of the carrier concentration which is a very noisy magnitude in EMC simulators. As a consequence, it is a hard task to obtain convergence and the corrections to the field should be calculated from different magnitudes trying to keep the accuracy of density gradient simulations.

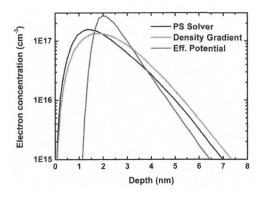

Fig. 2. Electron concentration profile obtained from the solution of the Schrödinger equation (black), the Density Gradient approach (red) and the effective potential (blue) for a MOS structure. As can be observed, the Density Gradient fairly reproduces the inversion charge profile [Reference 20].

2.1.3. *The effective conduction band edge (ECBE) method*

It is preferred, in MC simulations, to express the quantum correction in terms of the electrostatic potential, which is a smoother magnitude. The Effective Conduction Band-Edge method (ECBE) was developed to obtain the benefits of the density gradient in Monte Carlo simulations[28].

Starting from the density gradient, Eq (2), and assuming an exponential relation between the electron concentration and the effective potential, i.e.,

$$n \propto \exp\left(\frac{qV^*}{k_B T}\right) \tag{4}$$

the corrected potential V^* can be written as

$$V^* = V + V_q = V + \frac{q\hbar^2}{4 r m_n^* k_B T}\left(\nabla^2 V^* + \frac{q}{2k_B T}\left(\nabla V^*\right)^2\right) \tag{5}$$

The equation is solved in a self-consistent way with Poisson's equation and the BTE using the boundary conditions proposed by Jin et al.[29]. Depending on the device geometry (Figure 3-(left)) and bias point, the population of the different valleys may change. As a consequence, there is a variation in the confinement effective mass when only one valley is considered which has to be taken into account to reproduce the charge profile obtained from the solution of the 1D Poisson-Schrödinger system. The practical consequence is the use of the effective mass as a fitting parameter, as shown in Figure 3-(right). To avoid this issue, an improved version of the ECBE was developed to take into account the effect of different valleys and the orientation effects on the device performance.

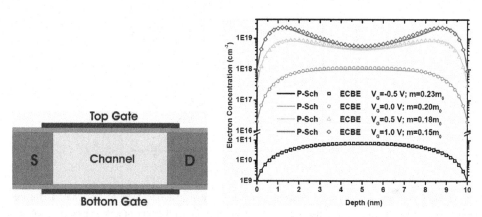

Fig. 3. (left) Double gate Silicon-on-insulator structure used in the simulations. (right) Poisson-ECBE and Poisson-Schrödinger solutions of the electron concentration in a 10 nm thick DGSOI structure for different gate bias. The ECBE approach gives a very good fit with the results obtained from the quantum calculations. However it is necessary to fit the effective mass for each bias condition.

2.1.4. *The multivalley effective conduction band edge approach (MV-ECBE)*

The standard ECBE method gives very good results when studying different SOI structures[22]. However, the main problem with this approach comes from the use of the effective mass as a fitting parameter. Nowadays, the study of SOI devices with orientations other than [100] (i.e. FinFETs) and the effects of carrier reorganization in the different valleys as the silicon thickness is reduced is of great interest. As a consequence, it is necessary to develop a new expression to take into account the electron population in different valleys and an arbitrary orientation in order to avoid using the effective mass as a fitting parameter. This new approach is called Multi-Valley ECBE (MV-ECBE)[22].

To obtain the MV-ECBE equation, let us consider as our starting point the one-particle Schrödinger equation representing the effective mass as a tensor

$$-\frac{\hbar^2}{2}\nabla\cdot\left[\left(\frac{\overset{\leftrightarrow}{1}}{m}\right)_j\nabla\Psi_j\right]+U\left(\vec{r}\right)\Psi_j=i\hbar\frac{\partial\Psi_j}{\partial t} \qquad (6)$$

where the sub-index j represents the valley. Assuming that the wave-function can be written as follows

$$\Psi_j=R_j\exp\left(\frac{iS_j}{\hbar}\right) \qquad (7)$$

where R_j and S_j are the modulus and phase of the solution respectively; if Equation (5) is substituted in Equation (4), and following the calculations in Reference 22, the correction term for the n-th valley can be written as follows:

$$V_j\simeq V+\frac{\hbar^2}{4qrV_T}\left\{\nabla\cdot\left[\left(\frac{\overset{\leftrightarrow}{1}}{m}\right)_j\cdot\nabla V_j^*\right]+\frac{1}{2V_T}\left[\nabla V_j^*\cdot\left(\frac{\overset{\leftrightarrow}{1}}{m}\right)_j\cdot\nabla V_j^*\right]\right\} \qquad (8)$$

which is the ECBE equation for a multi-valley system and arbitrary confinement directions. It is easy to demonstrate that equation (8) reduces to the standard density gradient theory, equation (5), when only one valley and a diagonal effective mass tensor with the same value for all the elements are assumed.

To demonstrate the performance of the MV-ECBE method, we have implemented it in a quantum-corrected Monte Carlo (QMC) simulator that includes new models to describe electron-phonon interaction and surface roughness scattering based on two-dimensional electron gas mechanisms. The starting point for the Q-EMC calculation is given by a 2D Poisson-Drift Diffusion solver which includes quantum corrections by the standard ECBE method. The use of a good initial guess reduces the time needed to reach stationary conditions. Since the Schrödinger equation is not solved in an explicit way, only one sub-band is considered for each valley. Self-consistency is preserved by updating the electrostatic potential and the quantum-corrected potential every *0.1fs* using the actual electron concentration given by Monte Carlo. The currents in the contacts are calculated by particle counting and Ramo-Shockley's theorem, described elsewhere[30].

The scattering mechanisms included in the simulations are acoustic intravalley phonons, intervalley phonons, Coulomb and surface roughness scattering. Intervalley scattering has been implemented using zero order approximation with three different phonons, following the work described in Reference 23. With the multi-valley approach, it is necessary to adapt the scattering expression to the case of a pseudo-2D electron gas in order to calculate the different valley populations in a self-consistent way. This is performed by including the calculation of overlapping factors between initial (i) and final (j) valleys

$$ F_{ij} = L_D \int_{T_{Si}} \left| \Psi_i(y) \right|^2 \left| \Psi_j(y) \right|^2 dy \tag{9} $$

where L_D is the Debye length, y the confinement direction and

$$ \left| \Psi_i(y) \right|^2 = \frac{n_i(y)}{\int_{T_{Si}} n_i(y) dy} \tag{10} $$

is the envelope of the electron wave function in the pseudo 2D gas approach. The scattering rates are then weighted by F_{ij}. In this way, size quantization is taken into account and the valley population is calculated self-consistently. Surface roughness scattering is implemented using a pseudo-2D gas version of the model proposed by Gamiz et al.[11], where an exponential spectrum model is assumed to represent the roughness at the silicon-oxide interface and the matrix elements depend on the overlapping integral between the perturbation potential and, once again, the envelope function of the corresponding valley. It is therefore very important to have a suitable representation of the envelope of the wave-function in each valley in order to calculate the scattering rates correctly. To test the performance of the simulator, different Double Gate SOI FETs (DGSOI) were considered. The chosen devices meet the requirements for the 22nm and 32nm nodes[1] with a channel length in the range of 10nm to 25nm. The silicon thickness (T_{Si}) varies from 6nm to 12nm, the effective oxide thickness (EOT) is 0.9nm and the metal gate contacts have a work-function value of 4.6eV. The drain and source doping concentration is $N_D=2\times10^{20}cm^{-3}$ and a very light doping concentration, $N_A=10^{15}cm^{-3}$, is used in the channel. Two surface orientations were considered for the DGSOI device, (100) and (110) with different options for the transport direction.

The first step, once the MV-ECBE is implemented in the Q-EMC code, is to check both the electrostatic and transport results, comparing these to well-established device simulators. Figure 4 (left) shows the actual carrier distribution when quantum corrections are taken into account at the midpoint between drain and source of the DGSOI transistor for the (100) surface orientation with $V_G = 1V$ and $V_{DS} = 0V$. As can be observed, the maximum of the distribution is no longer at the interface and its position corresponds to that predicted by the solution of the Schrödinger equation (solid line). The MV-ECBE solution is sketched with a dashed line and corresponds to the addition of the contributions due to each valley, represented by symbols. Valley 1 corresponds to the heaviest confinement mass, which is the most populated, as theory would lead us to expect, showing the breaking of the degeneracy of band minima compared to the three-dimensional electron gas model. The degeneration in Valleys 2 and 3 is also broken when a lateral electric field is applied. This is because electrons in such valleys have the same confinement mass but a different mass in the transport direction.

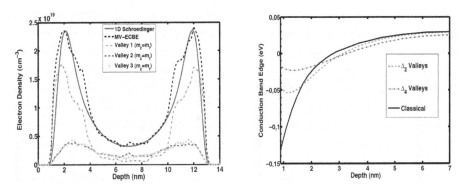

Fig. 4. (left) Cross-section of electron density in the center of the channel in the DGSOI device with $T_{ox} = 0.9nm$ when 1V is applied to each gate. Poisson-Schrödinger calculation (solid) is compared with MC results using the MV-ECBE (dashed). The contribution from each valley is also shown (symbols). (right) Detail of the electrostatic potential in a DGSOI MOSFET near the Si-SiO$_2$ interface when 1V is applied to both gates. Classical (solid) and quantum corrected potentials used to calculate the driving force for each valley in EMC simulations are represented. The repulsive potential can be observed near the interface (depth = 0.9nm) for MV-ECBE calculations.

Figure 4 (right) shows the classical Conduction Band Edge (solid) and the corrected band edge corresponding to each valley, again for the (100) surface. A repulsive term can be observed near the silicon-oxide interfaces (depth = 0.9 nm) shifting the centroid of the carrier distribution a certain distance from the silicon-oxide interface a certain distance, which coincides with the centroid position predicted by the quantum solution. The quantum correction tends towards zero where confinement effects are negligible. As can be observed, the MV-ECBE introduces different corrections for each valley since the confinement effective masses are not the same.

To calibrate the MV-ECBE simulator in terms of transport, a set of simulations were performed to calculate mobility curves in a long channel DGSOI device when a moderate lateral field was applied. The data were extracted by evaluating the following expression at the midpoint of the channel

$$\mu = \frac{\int_{T_{Si}} \frac{v_x(y)}{f_x(y)} n(y) dy}{\int_{T_{Si}} n(y) dy} \qquad (11)$$

where the integrals are calculated along the confinement direction and v_x and f_x respectively correspond to the local velocity and electric field components in the transport direction. Phonon, Coulomb and surface roughness scattering were considered, with $\Delta_{SR} = 0.185nm$ and $L_{SR} = 1.5nm$ for both interfaces. The electron mobility was evaluated with this Q-EMC and compared with the results obtained from a one-particle 1D MC which self-consistently solves the Poisson-Schrödinger equations set. This 1D MC simulator has been widely used in several previous studies and thoroughly explained elsewhere[31]. Figure 5-(left) shows the results of the comparison between MV-ECBE (symbols) and 1D-MC (solid line) mobility calculations. The agreement observed between the two results is very good.

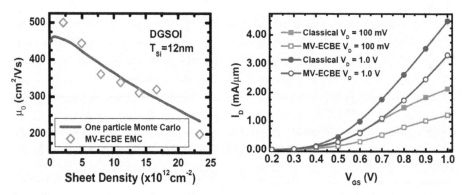

Fig. 5. (left) Comparison between mobility curves obtained from one-dimensional QMC (solid) and the Q EMC (symbols) for a 12nm thickness DGSOI. Phonon, Coulomb and surface roughness scattering are considered. (right) Classical (solid symbols) and quantum-corrected (open symbols) I-V curves comparison for the DGSOI MOSFET. When MV-ECBE is used, a degradation is observed in the sub-threshold slope. DIBL and threshold voltage are also increased.

A set of simulations was performed to compare the results obtained when classical and quantum-corrected approaches were considered. I-V calculations were performed to study the effect of quantum corrections on the transfer characteristics of the (100) DGSOI device. As shown in Figure 5-(right), the classical MC simulations clearly overestimate the current when compared with the quantum corrected results. Threshold voltages were calculated and a lower value was obtained in the classical simulation than in the QEMC (e.g. $V_{th-quantum} = 0.54V$ and $V_{th-classical} = 0.47V$ for $V_{DS} = 0.1V$). Sub-threshold swing was also evaluated, showing a degradation in the quantum-corrected case: *70mV/dec* for classical and *76mV/dec* for quantum with $V_{DS} = 1V$. Finally, drain-induced barrier lowering (DIBL) also increased when quantization was taken into account, its value varying from $DIBL_{classical} = 28mV/V$ to $DIBL_{quantum} = 34mV/V$. In conclusion, it can be observed that, as predicted by the theory, degradation in the performance of the device occurs due to quantum effects. Steady state situations were studied under $V_{GS} = V_{DS} = 1V$ bias conditions. In Figure 6 electrostatic potential distribution (top left) and electron density contour plots (top right) are shown for a (100) oriented DGSOI. Once more, the electron concentration profile can be observed, showing the effects of confinement in the channel. The device is in saturation as revealed by the pinch-off area. For the sake of comparison, Figure 6 (bottom) shows the inversion charge along the transport direction for the DGSOI device. As predicted by the Poisson-Schrödinger calculations, the inversion charge is smaller when quantum corrections are taken into account, especially in the channel area, where the differences between the two approaches are more significant. This effect increases the threshold voltage, as mentioned previously.

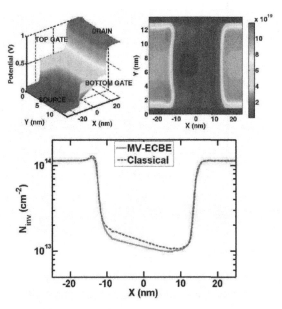

Fig. 6. Electrostatic potential (top left) and electron concentration (top right) plots of a DGSOI device when a symmetrical voltage of 1V is applied to both gates and the drain to source voltage is $V_{DS} = 1V$. Note how the maximum value of the electron concentration is again shifted from the oxide interface to the inside part of the channel. Inversion charge plot along the channel direction (bottom) for classical (dashed) and MV-ECBE (solid) simulations. The reduction in value for the quantum case implies an increase in the threshold voltage (Reference 17).

Thanks to the self-consistent adjustment of the valley population in the MV-ECBE method, certain quantum effects in ultra-thin body devices (UTB) arise naturally from the calculations. Figure 7 sketches the drain current dependence when T_{Si} is reduced and the inversion charge is kept constant. For small values of T_{Si}, an increase in the drain current is observed. This effect, known as volume inversion, is mainly due to the strong coupling between the two channels produced in UTB SOI transistors[32]. However, a marked decrease in the drain current is also observed as the silicon thickness is reduced below the maximum given by the volume inversion. This reduction is a direct consequence of quantum confinement. When the silicon thickness is reduced, the electrons are more confined and thus the uncertainty in the position decreases. This means that, as Heisenberg's principle predicts, uncertainty in the momentum increases. Its main effect is a significant increase in phonon scattering, which dramatically reduces the electron mobility. The effective conduction mass is also represented (inset), showing the inter-subband modulation effect. This produces a reduction of the effective mass when T_{Si} decreases because the population of the first valley, with the smallest conduction mass, increases.

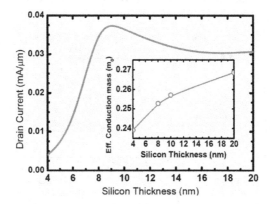

Fig. 7. Drain current and effective conduction mass (inset) versus channel thickness plots. Volume inversion and inter-subband Modulation effects can be observed.

2.2. *Multisubband-Ensemble Monte Carlo method*

The drawback of the MV-ECBE approach is the necessity of an *a-priori* calculated valley population. This fact limits the application of the model to drift diffusion simulations.

As mentioned before, a second option to face the problem of quantum confinement in EMC simulators is the use of the Multi-Subband method (MSB-EMC). The Multi-Subband method is based on the mode-space approach of quantum transport[33]. The quantum problem is considered as decoupled in the transport direction and, therefore, the Schrödinger equation has to be solved only in the confinement direction. The main limitation of the method is the impossibility of including coherence phenomena in a direct way that could be of special interest for the study of some structures. This is not the case for standard DGSOI devices. From the simulation point of view, our transistor is considered as a stack of slices perpendicular to the confinement direction z (Figure 8). The 1D Schrödinger equation for each slice and each valley is then solved self-consistently with the 2D Poisson equation. As a result the evolution of the eigen-energies, $E_{i,v}(x)$ and the wavefunctions, $\xi_{i,v}(x,z)$, is obtained along the transport axis, x, for the i-th valley and the v-th subband. The wave functions in the center of the channel, i.e. $x = L_{Ch}/2$, corresponding to the first three levels of un-primed valleys and the quantum well are represented for a (001) oriented device in Figure 9-(left). The transport from source to drain is considered as semiclassical and the Boltzmann Transport equation is solved by the MC method in the transport plane (Figure 8). Non parabolicity effects are included in both transport and solution of Schrödinger equation following Fischetti et al.[23]. In contrast to the standard MC semiclassical codes, the driving field expirienced by a simulated super-particle is not obtained from the gradient of the electrostatic potential.

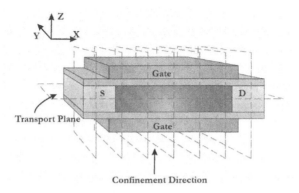

Fig. 8. Multisubband Ensemble Monte Carlo simulation of a DGSOI. 1D Schrödinger equation is solved for each grid point in the transport direction whereas Boltzmann Transport equation is solved by the Monte Carlo method in the transport plane.

According to the space-mode approach, the drift field is calculated from the derivative of $E_{i,v}(x)$, i.e. the driving force is different for each subband corresponding to a given valley. This can be inferred from Figure 9-(right) where the different evolution of the three first energy levels along the transport channel is represented for a 4 nm channel thick DGSOI.

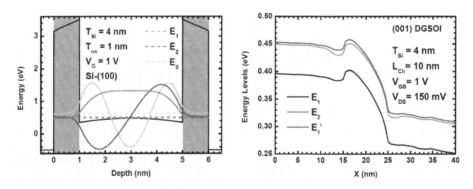

Fig. 9. (left) Quantum well, three first energy levels and normalized wave functions for a (100) oriented device in the center of the channel. Gray areas represent oxide layers. (right) Subband profile along the transport direction corresponding to the three first energy levels for a 10 nm length and 4 nm thick DGSOI device.

Concerning the calculation of scattering rates, a 2D electron gas is assumed since a full quantum description of the properties of the carriers is used for the confinement direction. Acoustic and intervalley phonon scattering rate calculations are detailed in Reference 23, whereas surface roughness scattering is described in Reference 11. All the models consider nonparabolic and ellipsoidal bands. As the quantum well used to solve the Schrödinger equation is different for each considered slice, the obtained eigen-energies and wavefunctions are also different. Thus, it is necessary to calculate a scattering table for each grid point, i.e. slice, and update them for every new solution of the Schrödinger equation needed to keep the self-consistency of the calculations. The subband population, $N_{i,v}(x)$, is calculated by resampling the super-particles belonging to a given subband and slice using the cloud-in-cell method[34]. The electron concentration, $n(x,z)$, is obtained adding all the probability densities $|\xi_{i,v}(x,z)|^2$ weighted by the corresponding subband population:

$$n(x,z) = \sum_{i,v} N_{i,v}(x)|\xi_{i,v}(x,z)|^2 \qquad (12)$$

The electrostatic potential is updated by solving the 2D Poisson equation using the previous $n(x,z)$ as input. This approach is especially appropriate for the study of 1D confinement in nanoscale devices. The drawback compared with semi-classical MC codes, is an important increase of the computational effort.

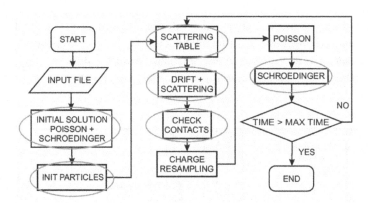

Fig. 10. Flowchart of the MSB-EMC simulator used in this work. Parallel routines are highlighted with orange circles.

This issue is partially overcome thanks to the parallel implementation of the code. For the sake of clarity, the flowchart of the simulator is shown in Figure 10 with the parallel routines highlighted with circles. More than a 90% of the code has been parallelized using OPEN-MP directives[35] achieving a speed-up of 7 for simulations performed using eight cores machines.

The MSB-EMC method has been used to study physical effects and it also could help to design new devices. We have used the simulator to study the orientation effects in double gate SOI MOSFETs, but also to see if the combination Ground-Plane+UTBOX could extend the scaling limits of FD-SOI devices.

2.3. *Multisubband-Ensemble Monte Carlo validation*

To validate the MSB-EMC simulator for ultrashort devices, a set of experimental data has been compared to our simulations. To perform a fair comparison it is necessary an in-depth knowledge of the technological parameters. A detailed enough description of ultrathin FDSOI devices is presented in Reference 36 where a SGSOI with T_{Si}= 8nm, L_G= 33nm, Ultrathin-BOX (10 nm) and ground plane (GP) is characterized. The gate stack is formed by TiN (10nm), HfO$_2$ (2.2 nm) and SiO$_2$ (1.6nm). Figure 11 shows I_D-V_{GS} curves for V_{DS}= *1.1V* (solid) and V_{DS}=*100mV* (dashed). Our simulations (symbols) show a good agreement with experimental data (lines).

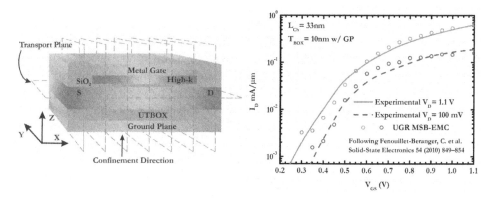

Fig. 11. (left) Simulated structure for the considered ET-FDSOI transistors. 1D Schrödinger equation is solved for each grid point in the transport direction whereas Boltzmann Transport equation is solved by the Monte Carlo method in the transport plane. (right) I_D vs. V_{GS} curves for a 33nm SGSOI with ground plane. Experimental results from Reference 36 (lines) are compared to the MS-EMC simulator (symbols). A good agreement is obtained for both linear and saturation regime (dashed and solid lines respectively).

Once the code was validated, we have used it to study hot-topics such as the scaling of FDSOI transistors, and the behavior of DGSOI devices with the crystallographic orientation.

3. Optimization of ultrathin fully-depleted SOI transistors with ultrathin buried oxide (BOX)

From the point of view of the control of short channel effects, ultrathin SOI technology presents another important advantage compared to standard bulk devices: the addition of new degrees of freedom in the definition of the device geometry, i.e., the thicknesses of the silicon and buried oxide layers. In this way, the scaling is carried out thanks to a channel thickness reduction instead of the implantation of complicated doping profiles with very high doping densities. Following the classical scaling rules for bulk MOSFETS, the maximum depth for source and drain implants must be close to $L_G/5$[37]. Therefore, the fabrication of decanano bulk devices demands an important reduction in the implant depth of source and drain regions with the subsequent challenge of obtaining higher doping densities stopping the dopant diffusion to reach the targeted depth. This issue disappears in ultra-thin SOI technology since the BOX constitutes a natural barrier to dopant diffusion. The scaling rules stand also different for bulk and SOI devices. In the case of FDSOI devices, the silicon thickness plays the role of the implant depth, however the conventional design rule can be modified to a less restrictive $L_G/T_{Si} = 4$[38]. In this way, a 4 nm silicon slab would be necessary for a 16 nm channel-length device. However, for T_{Si} smaller than 5 nm two effects limit the use of FDSOI devices: On one hand, it is very difficult to keep good enough thickness uniformity at wafer level to avoid V_{th} fluctuations. On the other hand, electron mobility is dramatically reduced as a consequence of confinement effects[39].

The condition $L_G/4$ can be relaxed for MuGFET devices where the recommended T_{Si} to minimize SCEs follows $N_G L_G/4$ being N_G the effective number of gates[40]. Therefore, if we use a double gate structure ($N_G=2$) the necessary thickness of the silicon slab in the previous example will be increased to 8nm; if we use a three-gates structure, the necessary silicon thickness will be 12nm. In both cases, the thickness of the silicon layer is large enough to avoid the above problems of degradation of the mobility and variability. In most cases, the use of multiple-gate devices means the use of 3D architectures whose mass production implies important efforts from an economical and technological point of view.

Therefore, it is necessary to relax the channel length to thickness constraint to extend the use of SGSOI transistors to the future nodes. There is still an unexploited way to improve SCEs control for SGSOI devices and, thus, to extend the use of FDSOI devices beyond the limit given by standard design rules considering a thick BOX, as it has been the standard up to now. The use of ultra thin BOX (UTBOX) and the addition of a ground plane (GP) can improve the behavior of the device. This fact has been already experimentally demonstrated for gate lengths of 33 nm as shown in Reference 36.

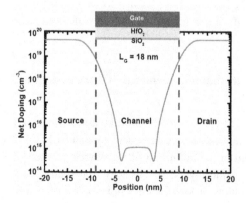

Fig. 12. Net doping profile for the 18 nm channel length device. It can be observed the residual doping of the channel and the gate stack structure is shown for the sake of clarity.

It is necessary then, to assess the possibilities that the combined use of UTBOX+GP offers to determine whether the scaling of FDSOI devices makes possible to fulfill the requirements of sub-32 nm nodes reducing the impact of thickness fluctuation effects. Hence, the rest of this Section is devoted to the study of such scaling possibilities and their impact on the performance of sub-32 nm ultrathin SOI devices. As the channel length is reduced, the control of DIBL (Drain Induced Barrier Lowering) and V_{th} roll-off is one of the biggest challenges from the point of view of device optimization. A set of FDSOI devices with channel lengths ranging from 32 to *11 nm* has been considered to assess the impact of different BOX configurations on SCEs and therefore on the scaling of sub-32 nm transistors. The gate stack structure includes a midgap metal and a HfO_2/SiO_2 dielectric bilayer with an EOT of *1.2nm* which is in the order of the recommended for a well tempered decanano device[1]. For all the cases T_{Si} is fixed to *6nm* which fits the requirements for a *24nm* channel length following the standard rule $L_G = 4T_{Si}$ [38]. Concerning the BOX, three configurations are studied:

(1) a standard thick BOX ($T_{BOX}=145$ nm),
(2) an Ultra-thin BOX (UTBOX, $T_{BOX} = 10$ nm)
(3) the same UTBOX ($T_{BOX} = 10$ nm) but including a ground plane contact (GP) with $N_A = 3\times10^{18} cm^{-3}$.

In all the cases, source and drain regions are doped with $N_D = 5.2\times10^{19} cm^{-3}$ where a Gaussian transition profile is considered into the channel with no variation of the doping in the transversal direction. The net doping profile is shown in Figure 12 for the 18 nm channel length device. It can be observed the residual doping in the channel, $N_A = 10^{15} cm^{-3}$ and the gate stack structure also sketched for the sake of clarity.

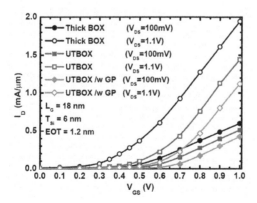

Fig. 13. I_D vs. V_{GS} curves for the 18 nm FDSOI devices calculated at $V_{DS} = 100mV$ (closed symbols) and $V_{DS}=1.1V$ (open symbols). It can be observed the important variation of the characteristics for the standard thick BOX devices demonstrating that the scaling for such architecture cannot be extended beyond the standard rules.

Due to the higher immunity to SCEs, a DGSOI obtained from the direct substitution of the BOX by an identical gate stack than in the FDSOI previously described, keeping the channel thickness to 6 nm, has been considered as reference device. In this case, and following the aforementioned scaling rules, the DGSOI transistor should be targeted for $L_G \simeq 4T_{Si} / N_G = 12nm$. As the proposed devices are in principle addressed for $L_G \simeq 24nm$, let us consider a structure beyond the scaling specifications for a standard SGSOI transistor to clearly check the impact of the three BOX configurations on transfer characteristics. Figure 13 shows I_D vs. V_{GS} for the 18 nm length FDSOI devices at $V_{DS}=100$ mV and $V_{DS} =1.1$ V (closed and open symbols respectively). As can be observed, the different BOX configurations have a considerable impact especially on the threshold voltage at both high and moderate drain bias. From the results it can be inferred that the use of thick BOX is not recommended due to the important variation on the V_{th} requiring a channel thinner than *5nm* in order to fulfill the scaling rules. However, this will not make things much better since the aforementioned problems from the point of view of performance and variability will start to play an important role. Focusing on UTBOX options, the variation of the characteristics from high to moderate V_{DS} remains under control and a closest study is necessary to determine whether UTBOX or UTBOX+GP could meet the minimum requirements on SCEs control.

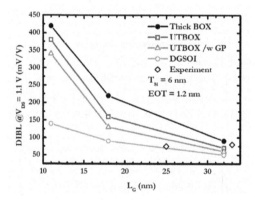

Fig. 14. DIBL as a function of the gate length including the different BOX configurations for FDSOI, and the corresponding DGSOI device (open circles). For the sake of comparison, experimental results obtained for L_G=33nm and L_G=25 nm obtained from Reference 36 and Reference 41 respectively are also shown (diamonds).

Figure 14 shows the channel length dependence of the Drain Induced Barrier Lowering (DIBL), defined at $V_{DS} = 1.1V$ as

$$DIBL \equiv \frac{\left(V_{TH @ V_{DS}=100mV} - V_{TH @ V_{DS}=1.1V}\right)}{\Delta V_{DS}} \tag{13}$$

for the different devices under study. As expected, there is an important increase of DIBL as L_G is reduced, however, the UTBOX+GP device keeps it under control (130mV/V) for the 18 nm device (triangles). The results corresponding to DGSOI devices with $T_{Si} =6$ nm show smaller values of DIBL as a consequence of the higher electrostatic control presented by these devices. Diamond symbols represent experimental results for 33 nm[36] and 25 nm[41] ultrathin-FDSOI devices. A good agreement is found with the simulations especially for the 33 nm device. It must be highlighted that an important dispersion on the results can be found showing the impact of the different technological processes. As an example, the 25 nm device corresponds to a highly optimized geometry including new stress techniques, being one of the lowest DIBL and V_{th} roll-off reported up to now. Therefore, it could be considered as a practical limit for DIBL optimization. Bulk devices have also been simulated following the templates proposed in Reference 42 extended to 18 nm. In this case (not shown in Figure 14), the DIBL is close to 350 mV/V which is out of the range of targeted values confirming the difficulty of extending bulk technology beyond the 32 nm node.

Concerning another important consequence of scaling, V_{TH} roll-off as a function of the gate length, Figure 15-(left) shows the evolution of threshold voltage as a function of L_G for the three considered BOX configurations. As the gate length is decreased, a reduction of V_{th} is observed for all the cases; however the smallest variation among the FDSOI devices occurs again for the UTBOX configurations (triangles and squares). The use of UTBOX also increases V_{TH} respect to the standard thick BOX case for a given length due to the electrostatic influence of the BOX. This fact allows keeping a single metal gate for both p and n devices, which is not possible if V_{th} becomes very small since the noise margins are dramatically reduced.

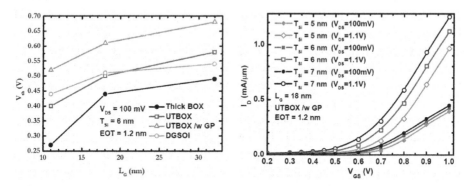

Fig. 15. (left) Threshold voltage as a function of the gate length for all the BOX configurations and the DGSOI reference device (open circles). For all the cases the same midgap metal gate is used. (right) I_D vs. V_{GS} curves for the 18 nm FDSOI devices calculated at $V_{DS}=100mV$ (closed symbols) and $V_{DS}=1.1V$ (open symbols) and T_{Si} ranging from *5nm* to *7nm*. The variation of the characteristics can be specially noticed for saturation bias conditions.

Finally, the impact of channel thickness has been also considered in the simulations. Figure 15-(right) shows I_D-V_{GS} curves for $V_{DS}=1.1V$ (open symbols) and $V_{DS}=100mV$ (closed symbols) corresponding to the 18 nm device and silicon thickness ranging from 5 to 7 nm. As observed, there is a V_{th} increase as the T_{Si} is reduced especially for the case of saturation bias conditions. As a consequence, there is an important variation in the DIBL value specially between the $T_{Si}=6nm$ and the $T_{Si}=7nm$ thick devices: $DIBL(@T_{Si}=6nm)=130mV/V$ and $DIBL(@T_{Si}=7nm)=210mV/V$. However there is a small variation (only 10 mV/V) when the channel thickness is reduced to 5 nm: $DIBL(@T_{Si}=5nm)=120mV/V$. Therefore, there could be an important influence of the thickness fluctuations on the SCEs control that cannot be neglected when extremely thin film devices are used.

All these features obtained from the use of FDSOI devices combined with UTBOX+ GP allow extending the use of SGSOI transistors for sub-32 nm nodes. The simulations show that 18 nm gate length devices with a channel thickness of 6 nm present good performance and excellent SCEs control. Therefore, the standard design rule which relates gate length and channel thickness can be relaxed to $L_G \simeq 3T_{Si}$ giving an extra technological node for a given T_{Si} and delaying the aforementioned end of the scaling capabilities based on channel thickness. Following the considerations presented in Reference 40, the proposed design rule corresponds to $N_G \simeq 1.5$, giving an idea of the extra electrostatic control obtained from the UTBOX+GP which could be represented as an additional *half-gate*. The main advantage of this configuration in between of single and double gate structures is the compatibility with the standard SGSOI fabrication flow. However, further studies are necessary in order to evaluate the impact of channel thickness variability on the performance of sub-32 nm node devices.

In summary, we have studied the impact of different BOX configurations on the scaling of FDSOI devices to sub-32 nm nodes. The scaling rules for FDSOI have been assessed up to now on the channel thickness, however the issues expected from both fabrication and device performance when $T_{Si}<5nm$ require new scaling boosters to relax the well known thumb-rule of $L_G \simeq 4T_{Si}$. Standard thick BOX, UTBOX and UTBOX+GP solutions have been considered in order to check their influence on SCEs. The simulations show that one of the main limiting factors for downscaling is the DIBL which is affected in an important manner by a reduction on the electrostatic integrity. UTBOX+GP configuration is the only valid option to downscale SGSOI beyond 20 nm channel length keeping the silicon slab above the theoretical limit of 5 nm, where thickness variability and mobility reduction will play an important role. Moreover, the use of UTBOX+GP allows us to redefine scaling rules, relaxing the L_G/T_{Si} ratio constraint to 3. In this way, an extra technological node could be achieved for a given channel thickness making FDSOI technology a feasible option to MuGFETs down to the 11 nm node.

4. Orientation effects in ultra-short channel DGSOI devices

As mentioned above multi-gate field effect transistors (MuGFETs) are among the preferred candidates to achieve the requirements of the International Technology Roadmap for the Semiconductor Industry (ITRS)[1] with regard to scaling beyond 20nm node. However, under the common MuGFET denomination, there is a wide variety of devices and configurations. Moreover, the possibility of fabricating both planar and vertical devices on standard SOI wafers allows different channel and confinement directions by rotating the device[43]. In the case of the most commonly used Si-(100) wafers, planar DGSOI transistors present quantum confinement perpendicular to (100) plane and the channel can be oriented on any of the $\langle 0\alpha\beta \rangle$ directions, being the standard one $\langle 011 \rangle$. When vertical double gate structures are used (i.e. FinFETs) different confinement directions can also be obtained. For standard designs, the FinFET is oriented parallel or perpendicular to the wafer flat. The channel is parallel to the (110) plane. A rotation of 45° on the transistor layout recovers a (100) device similar to the planar case. Intermediate rotations can be used to mimic the behavior of different orientations including (111)[43]. Thus, different device orientations can be easily included in a single layout. This fact has a special relevance in CMOS designs since measured hole mobilities show an improvement when the device is oriented in directions other than (100)[44] and a degradation for the case of electrons as presented in Reference 45. This solution can be a better and cheaper choice to maximize both electron and hole mobilities than those based on more exotic and expensive substrate solutions such as Hybrid-Orientation Technology (HOT)[46]. However, low-field mobility measurements do not take into account source/drain areas effects on the carrier transport and make the assumption of constant drift field which is not the case of ultra-short channel devices. A thorough study of such effects is also needed. A quantization direction different from the standard (100) modifies both electrons and holes mobilities due to the anisotropy of the effective masses in the silicon lattice[47,48].

Table 1. Numerical values of effective masses for electrons in silicon for different surface orientations

Surface Orientation	m_x	m_y	m_z	n_v
(100)	$0.190m_0$	$0.190m_0$	$0.916m_0$	2
	$0.190m_0$	$0.916m_0$	$0.190m_0$	4
(110)	$0.190m_0$	$0.553m_0$	$0.315m_0$	4
	$0.190m_0$	$0.916m_0$	$0.190m_0$	2
(111)	$0.190m_0$	$0.674m_0$	$0.258m_0$	6

In the case of the conduction band, silicon presents six constant energy ellipsoids as shown in Figure 16-(top). The effective masses corresponding to the main directions for a given ellipsoid are $m_l = 0.916\ m_0$ and $m_t = 0.19m_0$ where the sub-indexes l and t stand for longitudinal and transverse effective masses respectively. The values of the masses for the most used surface orientations are presented in Table 1. The confinement direction is considered as the Z coordinate, thus m_z is the quantization mass. The transport plane is given by the X and Y coordinates, being m_x and m_y the associated transport masses and n_v is the degeneracy of each set of valleys. The first row for each orientation corresponds to the set of valleys with the highest quantization mass. For the (100) case, two sets of subbands are considered. Non-primed subbands with $m_z = m_l$ and primed subbands with $m_z = m_t$. As $m_t < m_l$, non-primed subbands have lower quantization energy and transport mass. When (110) confinement direction is considered, the degeneracy of the valleys with the highest confinement mass is four. Finally, in the (111) oriented wafer, all the six ellipsoids have the same confinement mass and, therefore, there is only one set of subbands. The main consequence of the existence of different sets of subbands, is a dependence of the transport properties with the crystallographic orientation. This is due to the fact that carrier distribution, wave functions, energy levels, form factors and, therefore, scattering rates are affected by the confinement direction.

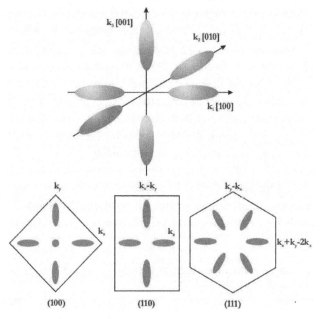

Fig. 16. (top) 3D representation of constant energy ellipsoids for Si conduction band. (bottom) Two dimensional projection of constant energy ellipses for (100), (110) and (111) Si conduction band. Lower energy subbands are shown in red.

Figure 16-(bottom) also suggests an anisotropy in the effective mass once the confinement direction is fixed. This fact also leads to variations on the performance of FinFETs depending on the channel orientation. This is especially important for the optimization of CMOS based circuits since the best device orientation for electron and hole based transistors is different[49]. A detailed study of orientation effects has to be carried out including not only mobility issues, but also the effect of source/drain regions on general performance of next-generation ultra-short channel MuGFETs. In previous works we have used a one-particle Monte Carlo method to study the effect of the crystallographic orientation in Single Gate and Double-Gate MOSFETs[13,45,50]. However, in today's scenario where manufacturing costs grow exponentially as device dimensions are aggressively scaled down, device simulators must be used to forecast the performance of future structures and decide which ones would be better for circuit applications.

Therefore, in order to reduce design costs, great efforts should be made in a priori device optimization.

4.1. *DGSOI drain current dependence on crystallographic orientation*

For the study of the impact of crystallographic orientation on the performance of ultra-short channel DGSOI nMOSFETs, the proposed structure is a device with an undoped channel of 10 nm length and 4 nm Si thickness. Midgap-metal gate contacts have been used with an equivalent oxide thickness (EOT) of *1nm*. To do so, the MSB-EMC simulator previously presented has been used. The increase of the computational time compared to other approaches based on semiclassical or quantum corrected codes is about 8-fold. However, the results obtained using such codes clearly over-estimate the drain current especially for moderate-high drain bias conditions. Figure 17 shows the output characteristics for the proposed device calculated with the semiclassical approach (black), the MV-ECBE (blue) and MSB-EMC (red). In the semiclassical simulation, drain current is higher since inversion charge is bigger than in MSB-EMC, confinement does not affect in a relevant way the device performance and the surface roughness scattering model is based on diffusive reflections. In the case of MV-ECBE approach, since only one subband per valley is considered, inter-valley and surface roughness scattering are underestimated. Therefore, it is necessary to use the Multi-Subband approach to correctly reproduce the behavior of such ultra-thin devices.

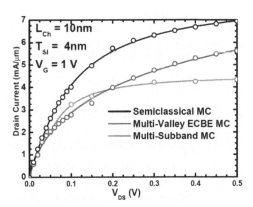

Fig. 17. Drain current comparison for different EMC approaches. Semiclassical (black) and MV-ECBE (blue) codes clearly over-estimate the value obtained with the MSB-EMC method (red).

To study the impact of device orientation, first of all, it is necessary to compare the electrostatic of the different options. The main effects are a redistribution of the energy levels and a change in the charge profile. Figure 18-(left) shows the electron distribution for a $T_{Si}=4nm$ thick channel considering the most commonly used confinement orientations for $V_G = 1$ V. The corresponding inversion charge is close to $1.4 \times 10^{13} cm^{-2}$ for all the cases. As observed, the carrier profiles for (110) and (111) devices are very similar while there is a significant difference with the charge profile obtained for the (100) device.

This fact can be explained taking into account the confinement mass, m_z, values presented in *Table 1*. In the (111) and (110) cases, m_z is similar and smaller than for the (100) device. This implies lower energy subbands in Si(100) for the 2-fold valleys which are the most populated for ultra-thin layers or high inversion conditions[8]. As a consequence, electrons are closer to the Si-SiO$_2$ interfaces and two different peaks can be observed whereas in the other situations volume inversion profiles appear more clearly[32]. In all the calculations, the penetration of wave functions in the oxide is taken into account as shows the non-zero electron concentration in the oxide areas (gray zones in Figure 18-left).

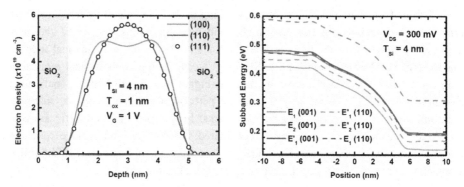

Fig. 18. (left) Electron concentration profile for different device orientations. Due to their similar confinement effective mass, (110) (blue) and (111) (black) structures present similar charge profiles. The higher effective mass for (100) orientation (red) leads to a higher confinement to Si-SiO2 interfaces. Gray areas represent oxide layers. (right) Three first eigen-energy profiles for (100) and (110) devices (solid and dashed lines respectively) when 300 mV are applied too drain contact.

It has been already commented that the variations in the confinement effective mass induce changes in the electron confinement and subband distribution[51]. As a consequence, the form factor value obtained from overlapping integrals which is used to calculate the scattering rate also varies and therefore, there is a direct effect on the transport properties and not only on the electrostatics. The variation of the eigen-energies with the channel direction also implies different energy profiles along the source/drain axis as shown in Figure 18-(right). The three first subbands are represented for the (100) and the (110) oriented devices when a drain voltage of 300 mV is applied. As can be observed, the set of subbands corresponding to the ground state can also change with the wafer orientation, e.g. the fundamental energy level corresponds to the 2-fold valleys in (001) and to the 4-fold valleys in (110) oriented devices. Two are the direct consequences of the variation in the subband profile:

(1) The different distance between energy levels for each orientation. The inter-subband scattering processes are affected in an important manner by this fact. If the total

energy of the carrier including the absorbed/emitted phonon is higher than the corresponding to the final subband, the scattering process may take place. In other case, it cannot happen. Therefore, the subband redistribution changes the scattering probability. As Figure 18-(right) shows, when the confinement direction is (110), subbands are closer each other than in (100) case and the inter-subband scattering increases. Thus, the impact of phonon scattering should be higher in (110) oriented devices.

(2) The change in the driving force calculated from the derivative of the energy profile as mentioned before.

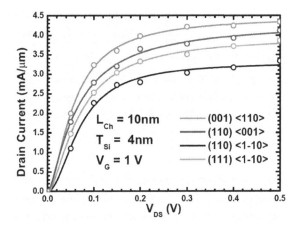

Fig. 19. Drain current for different channel orientations. Highest drain current is obtained for (100)/<110> (red). Transport mass anisotropy is shown for (110) results. Note the important differences for the <110> (black) and the <001> (blue) oriented channels.

The differences in the driving field are not very large and small variations in the drain current are expected. However, experimental results[49] and the simulations performed in this work show a stronger dependence. Figure 19 shows I_D vs V_{DS} curves for the main orientations for the 10 nm length device under study. The best performance is obtained for the (001)/<110> and the (110)/<001> devices. Since both transistors present similar inversion charge values, there is a direct extrapolation from velocity/mobility results to output characteristics[45]. The variation in the current values cannot be explained only by the changes in the driving field, especially the important differences in the drain current for (110) devices depending on the channel direction. This behavior is only observed in this wafer orientation, even when the drift field does not change in a relevant way if the channel orientation changes while the confinement direction is fixed. To explain this behavior, several factors have to be taken into account. The first one is the dependence of the transport effective mass on the channel orientation. In this way, drift velocity profiles change for each case as can be observed in Figure 20-(left). Higher velocities are observed for orientations with small transport mass, i.e. (001)/<110> and (110)/<001>.

The channel orientation dependence can be clearly observed for the (110) devices. The device faced to <001> shows a velocity profile close to the (001)/<110> with the best performance.

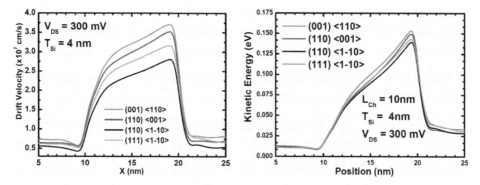

Fig. 20. (left) Drift velocity profile for the considered channel orientations with $V_{DS} = 300mV$. The highest drift velocity is obtained for the (100)/<110> (red) device. Drain current differences for (110) devices comes from transport mass anisotropy which produces important drift velocity variations for the <110> (black) and the <001> (blue) oriented channels. (right) Kinetic energy profiles along the channel for the devices under study for $V_{DS} = 300mV$.

However, a 90-degrees rotation of the channel, i.e. <1-10>, degrades the drift velocity leading to the poorest performance. In the case of (111) and (001) orientations, there is no dependence of the performance with the channel orientation since valley ellipsoids are symmetrically placed.

A second effect that modulates the dependence with the transport effective mass has to be also taken into account. If the kinetic energy profiles are also studied, as in Figure 20-(right), it can be observed how the differences are smaller than in the drift velocity plot. The increase in the effective mass for some orientations can explain in part the stronger dependence of the velocity comparing to the energy case. However scattering plays also a role that cannot be neglected. As long as the drift velocity decreases, the transit time along the channel defined as

$$\tau_t = \int_{L_{ch}} \frac{1}{v_{\parallel}(x)} dx \qquad (12)$$

increases. This means that carriers spend more time in the channel. Besides, the average energy and, therefore, the scattering rates are of the same order, and therefore, each carrier suffers a greater amount of scattering events. As a consequence, the drift velocity is reduced more than expected. This explains how the difference among the velocities increases as the carriers approach the end of the channel (Figure 20-left).

In summary, we have used a Multi-Subband Ensemble Monte Carlo device simulator to study the impact on the performance of ultra-short channel DGSOI transistors including different confinement and transport orientations. Our simulations show high drift velocity and drain current for (001) and (110)/<001> devices. However, channel rotations can imply important degradations of the characteristics, as in the (110)/<1-10> case. The variations in the device performance are due to the change in the confinement and transport effective masses and the redistribution of the electron among the different subbands which imply an increase in the inter-subband scattering for non-(100) devices.

Acknowledgments

The authors would like to thank the support given by the European Union (EUROSOI+ CA and NANOSIL), the Spanish Government (FIS2008-05805, TEC2008-06758-C02-01), *Ramón y Cajal* program and Junta de Andalucia (P06-TIC1899 and P09-TIC4873).

References

1. International Technology Roadmap for the Semiconductor Industry, "http://www.itrs.net/."
2. C. Millar, D. Reid, G. Roy, S. Roy, and A. Asenov, "Accurate Statistical Description of Random Dopant-Induced Threshold Voltage Variability" Electron Device Letters, IEEE, vol. 29,pp. 946-948, (2008).
3. F. J. G. Ruiz, A. Godoy, F. Gamiz, C. Sampedro, and L. Donetti, "A comprehensive study of the corner effects in Pi-gate MOSFETs including quantum effects", Electron Devices, IEEE Transactions on, vol. 54, pp. 3369-3377 (2007).
4. O.Faynot, F.Andrieu, O.Weber, C.Fenouillet-Béranger, P.Perreau, J.Mazurier, T.Benoist, O.Rozeau, T.Poiroux, M.Vinet, L.Grenouillet, J-P.Noel, N.Posseme, S.Barnola, F.Martin, C.Lapeyre, M.Cassé, X.Garros, M-A.Jaud, O.Thomas, G.Cibrario, L.Tosti, L.Brévard, C.Tabone, P.Gaud, S.Barraud, T.Ernst and S.Deleonibus, "Planar Fully Depleted SOI Technology: a powerful architecture for the 20nm node and beyond", IEEE International Electron Device Meeting, 2010, pp.3.2.1-3.2.4 (2010)
5. G. K. Celler and S. Cristoloveanu, "Frontiers of Silicon-on-Insulator," J. Appl. Phys., vol. 93, pp. 4955-4978, (2003).
6. A. Khakifirooz, K. Cheng, B. Jagannathan, P. Kulkarni, J. Sleight, D. Shahrjerdi, J. Chang,S. Lee, J. Li, H. Bu, R. Gauthier, B. Doris, and G. Shahidi, "Fully depleted extremely thin SOI for mainstream 20nm low-power technology and beyond," in Solid-State Circuits Conference Digest of Technical Papers (ISSCC), 2010 IEEE International, pp.152-153, (2010)
7. S.Cristoloveanu and S.S.Li, Electrical Characterization of Silicon-on- Insulator Materials and Devices (Kluwer, Boston, 1995)
8. F.Gamiz, M.V.Fischetti, "Monte Carlo simulation of double-gate silicon-on-insulator inversion layers: The role of volume inversion" J Appl Phys 89, pp-5487-5478 (2001)
9. F.Gámiz, J.B. Roldán, and J.A. López-Villanueva, "Phonon-limited electron mobility in ultrathin silicon-on-insulator inversion layers" J.Appl.Phys. 83, pp-4802-4806 (1998)
10. F.Gámiz, J.B.Roldán, P.Cartujo-Cassinello, J.E.Carceller, J.A.López-Villanueva, S.Rodriguez, "Electron mobility in extremely thin single-gate silicon-on-insulator inversion layers", J.Appl.Phys. 86, pp.6269-6275 (1999)
11. F.Gámiz, J.B.Roldán, J.A.López-Villanueva, P.Cartujo-Cassinello, J.E.Carceller, "Surface roughness at the Si-SiO$_2$ interfaces in fully depleted silicon-on-insulator inversion layers", J.Appl.Phys., 6854-6863 (1999)
12. L.Ge, F Gámiz,G.O. Workman, S.Veeraraghavan, "On the gate capacitance limits of nanoscale DG and FD SOI MOSFETs" IEEE Trans.on Elec.Dev., 53, pp.753-758 (2006)

13. L.Donetti, F.Gamiz, "Hole transport in DGSOI devices: Orientation and silicon thickness effects", Solid State Electronics, 54, 191-195 (2010)

14. N.Bannov, V.Mitin, M.Stroscio, "Confined acoustic phonons in a free-standing quan-tum well and their interaction with electrons", Phys Stat Sol (B) 183, pp.131-135 (1994)

15. L.Donetti, F.Gamiz, J.B.Roldán, and A.Godoy, "Acoustic phonon confinement in silicon nanolayers: Effect on electron mobility" J.Appl.Phys, 100, 013701 (2006)

16. L.Donetti, F.Gamiz, N.Rodriguez, F.Jimenez-Molinos, C.Sampedro, "Influence of acoustic phonon confinement on electron mobility in ultrathin silicon on insulator layers" Appl.Phys.Lett. 88, 122108 (2006)

17. C. Sampedro, F.Gamiz, A. Godoy, F.Jimenez-Molinos, "Quantum Ensemble Monte Carlo simulation of silicon-based nanodevices", J. Comp.Elec, 6, pp.41-44, (2007)

18. U. Ravaioli and T. A. V. der Straaten, Simulation of Nanoscale Molecular Systems. Handbook of Theoretical and Computational Nanotechnology. American Scientific Publishers, 2006

19. R. Ravishankar, G. Kathawala, U. Ravaioli, S. Hasan, and M. Lundstrom, "Comparison of Monte Carlo and NEGF Simulation of Double Gate MOSFETs" J. Comp. Elec.,4 , pp.39-43, (2005).

20. A. Asenov, A. R. Brown, and J. R. Watling, "Quantum Corrections in the Simulation of Decanano MOSFETs," Solid State Elec., vol. 47, pp.1141-1145 (2003).

21. D. Ferry, S. Ramey, L. Shifren, and R. Akis, "The Effective Potential in Device Modeling: the Good, the Bad and the Ugly," J. Comp. Elec.,1, pp.59-65, (2002).

22. C. Sampedro-Matarin, F.Gamiz, A.Godoy, and F.Garcia-Ruiz, "The Multivalley Effective Conduction Band-Edge Method for Monte Carlo Simulation of Nanoscale Structures," Electron Devices, IEEE Trans.Elec.Dev., 53, pp.2703-2710 (2006).

23. M. V. Fischetti and S. E. Laux, "Monte Carlo Study of Electron Transport in Silicon Inversion Layers", Physical Rev. B, 48, no. 4, pp. 2244-2274 (1993).

24. J. Saint-Martin, A. Bournel, F. Monsef, C. Chassat, and P. Dollfus, "Multi Sub-Band Monte Carlo Simulation of an Ultra-Thin Double Gate MOSFET with 2D Electron Gas", Semicond.Sci. and Tech., 21, pp. 29-31, (2006).

25. I. Riolino, M. Braccioli, L. Lucci, D. Esseni, C. Fiegna, P. Palestri, and L. Selmi, "Monte-Carlo Simulation of Decananometric Double-Gate SOI devices: Multi-Subband vs. 3D-Electron Gas with Quantum Corrections", in Proceedings of the 36th European Solid-State Device Research Conference, 2006. ESSDERC 2006.

26. E. Sangiorgi, P. Palestri, D. Esseni, C. Fiegna, and L. Selmi, "The Monte Carlo Approach to Transport Modeling in Deca-Nanometer MOSFETs," Solid-State Electronics, 52, pp.1414-1423 (2008).

27. M.G. Ancona, G.J. Iafrate, "Quantum correction to the equation of state of an electron gas in a semiconductor", Phys. Rev. B 39(13), pp.9536-9540 (1989)

28. T.W.Tang and B.Wu," Quantum correction for the Monte Carlo simulation via the effective conduction-band edge equation" Semicond. Sci. Technol. 19, pp.54-60 (2004).

29. S.Jin, Y.Park, H.Min," Simulation of quantum effects in the nano-scale semiconductor device" Semicond. Technol. and Science 4, pp.32-40 (2004)

30. S.Babiker, A.Asenov, N.Cameron, S.P .Beaumont, and J.R.Barker, "Complete Monte Carlo RF analysis of "real" short-channel compound FET's", IEEE Trans. Elec. Dev. 45, pp.1644-1652 (1998).

31. F.Gamiz, J.Lopez-Villanueva, J.Banqueri, J.Carceller, and P.Cartujo, "Universality of electron mobility curves in MOSFETs: a Monte Carlo study", IEEE Trans.Elec.Dev. 42, pp.258-265 (1995).

32. F. Balestra, S. Cristoloveanu, M. Benachir, J. Brini, and T. Elewa, "Double-gate Silicon-On-Insulator Transistor with Volume Inversion: A New Device with Greatly Enhanced Performance," IEEE Elec. Dev. Lett., 8, pp. 410-412, (1987)

33. R. Venugopal, Z. Ren, S. Datta, M. S. Lundstrom, and D. Jovanovic, "Simulating Quantum Transport in Nanoscale Transistors: Real Versus Mode-Space Approaches", J.Appl.Phys, 92, pp. 3730-3739, (2002)

34. K. Tomizawa, in Numerical Simulation of Submicron Semiconductor Devices (Artech House, Boston, 1993)

35. The OpenMP Specifications for Parallel Programming, http://www.openmp.org, 2008

36. C.Fenouillet-Beranger, P.Perreau, S.Denorme, L.Tosti, F.Andrieu, O.Weber, S.Monfray, S.Barnola, C.Arvet, Y.Campidelli, S.Haendler, R.Beneyton, C.Perrot, C.de Buttet, P.Gros, L.Pham-Nguyen, F.Leverd, P.Gouraud, F.Abbate, F.Baron, A.Torres, C. Laviron, L.Pinzelli, J.Vetier, C.Borowiak, A.Margain, D.Delprat, F.Boedt, K.Bourdelle, B.Y.Nguyen, O.Faynot, and T.Skotnicki, "Impact of a 10 nm ultra-thin BOX (UTBOX) and ground plane on FDSOI devices for 32 nm node and below", Solid-State Electronics, 54, pp. 849-854, (2010).

37. R. Dennard, F. Gaensslen, V. Rideout, E. Bassous, and A. LeBlanc, "Design of ion-implanted MOSFET's with very small physical dimensions", IEEE J. Solid-State Circuits,9, pp.256-268 (1974)

38. J.-P. Colinge, in *FinFETs and Other Multi-Gate Transistors*, Springer Publishing Company, 2007.

39. F. Gamiz, J. B. Roldan, J. A. Lopez-Villanueva, P. Cartujo-Cassinello, and F. Jimenez-Molinos, "Monte Carlo simulation of electron mobility in silicon-on-insulator structures", Solid-State Electronics, 46, pp.1715-1721 (2002)

40. S. Cristoloveanu, "How Many Gates do we Need in a Transistor?", in Proceedings of International Semiconductor Conference, 2007. CAS 2007 (Sinaia, Romania), pp.3-10, (2007)

41. K. Cheng, A. Khakifirooz, P. Kulkarni, S. Ponoth, J. Kuss, D. Shahrjerdi, L. Edge, A. Kimball,S. Kanakasabapathy, K. Xiu, S. Schmitz, A. Reznicek, T. Adam, H. He, N. Loubet, S. Holmes,S. Mehta, D. Yang, A. Upham, S.-C. Seo, J. Herman, R. Johnson, Y. Zhu, P. Jamison,B. Haran, Z. Zhu, L. Vanamurth, S. Fan, D. Horak, H. Bu, P. Oldiges, D. Sadana, P. Kozlowski,D. McHerron, J. O'Neill, and B. Doris, "Extremely thin SOI (ETSOI) CMOS with record low variability for low power system-on-chip applications", in Electron Devices Meeting (IEDM), 2009 IEEE International, pp.1-4, (2009)

42. P.Palestri, C.Alexander, A.Asenov, V.Aubry-Fortuna, G.Baccarani, A.Bournel, M.Braccioli, B.Cheng, P.Dollfus, A.Esposito, D.Esseni, C.Fenouillet-Beranger, C.Fiegna, G.Fiori, A.Ghetti, G.Iannaccone, A.Martinez, B.Majkusiak, S.Monfray, V.Peikert, S.Reggiani, C.Riddet, J.Saint-Martin, E.Sangiorgi, A.Schenk, L.Selmi, L.Silvestri, P.Toniutti, and J.Walczak, "A comparison of advanced transport models for the computation of the drain current in nanoscale nMOSFETs" Solid-State Electronics, 53, pp.1293-1302, (2009)

43. L. Chang, M. Ieong, and M. Yang, "CMOS Circuit Performance Enhancement by Surface Orientation Optimization", IEEE Trans.Elec.Dev., 51, pp.1621-1627 (2004)

44. M. Yang, E. Gusev, M. Ieong, O. Gluschenkov, D. Boyd, K. Chan, P. Kozlowski, C. D'Emic, R. Sicina, P. Jamison, and A. Chou, "Performance Dependence of CMOS on Silicon Substrate Orientation for Ultrathin Oxynitride and HfO_2 Gate Dielectrics," IEE· Elec.Dev.Lett, 24, pp.339-341, (2003)

45. F. Gamiz, L. Donetti, and N. Rodriguez, "Anisotropy of Electron Mobility in Arbitrarily Oriented FinFETs," in Proceedings of the 37th European Solid State Device Research Conference, ESSDERC, 2007.

46. M. Yang, V. Chan, K. Chan, L. Shi, D. Fried, J. Stathis, A. Chou, E. Gusev, J. Ott, L. Burns, M. Fischetti, and M. Ieong, "Hybrid-Orientation Technology (HOT): Opportunities and Challenges", IEEE Trans.Elec.Dev., 53, pp.965-978, (2006).

47. T.Sato, Y. Takeishi, H.Hara, and Y.Okamoto, "Mobility Anisotropy of Electrons in Inversion Layers on Oxidized Silicon Surfaces," Phys. Rev. B, 4, pp.1950-1960, (1971)

48. F. Stern and W. E. Howard, "Properties of the Semiconductor Surface Inversion Layer in the Electric Quantum Limit", Phys. Review,163, pp.816-835, (1970).
49. G.Tsutsui, T.Hiramoto, "Mobility and Threshold-Voltage Comparison Between (110)- and (100)-Oriented Ultrathin-Body Silicon MOSFETs", IEEE Trans.Elec.Dev. 53, pp.2582-2588 (2006)
50. L.Donetti, F.Gamiz, N.Rodriguez and A.Godoy, "Hole Mobility in Ultrathin Double-Gate SOI Devices: The Effect of Acoustic Phonon Confinement", IEEE Elec.Dev.Lett., 30, pp.1338-1340 (2009)
51. C.Sampedro, F.Gámiz, A.Godoy, R.Valín, A.García-Loureiro, F.G. Ruiz, "Multi-Subband Monte Carlo study of device orientation effects in ultra-short channel DGSOI", Solid State Electronics, 54, pp.131-136 (2010)

ANALYTICAL MODELS AND ELECTRICAL CHARACTERISATION OF ADVANCED MOSFETs IN THE QUASI BALLISTIC REGIME

RAPHAEL CLERC

Laboratoire Hubert Curien (UMR 5516 CNRS)
Institut d'Optique & Université Jean-Monnet
Rue du Professeur Benoit Lauras, 42000, Saint-Etienne, France
raphael.clerc@institutoptique.fr

GERARD GHIBAUDO

IMEP LAHC (UMR 5130 CNRS)
Grenoble INP, Minatec
3, rue Parvis Louis Néel, 38000 Grenoble, France
ghibaudo@minatec.inpg.fr

The quasi-ballistic nature of transport in end of the roadmap MOSFETs device is expected to lead to significant on state current enhancement. The current understanding of such mechanism of transport is carefully reviewed in this chapter, underlining the derivation and limits of corresponding analytical models. In a second part, different strategies to compare these models to experiments are discussed, trying to estimate the "degree of ballisticity" achieved in advanced technologies.

Keywords: Advanced MOSFETs; Quasi Ballistic transport; Electrical Characterization; Neutral Defects.

1. Introduction

Since 2003, the ITRS roadmap has considered the Quasi Ballistic (QB) regime of transport as a possible "technological booster" of MOSFET performances[1]. Indeed, the physics of quasi ballistic transport was expected to lead to enhanced on state drive current I_{on}, compared to prediction based on the conventional drift diffusion theory. At very short channel length, as illustrated on Figure 1, the commonly used drift diffusion theory[2] predicts a saturation of the on current versus channel length, due to the mechanism of velocity saturation (v_{sat})[3][4][5]. The drift diffusion theory is based on a low field simplification of the semiclassical Boltzmann Transport Equation[2] and empirically accounts for the phenomenon of saturation velocity observed in long samples at high field condition, by introducing a longitudinal field dependent mobility equation[5][6]. This approach has been successfully applied to model relatively long device, but it does not apply at channel length comparable or lower than the mean free path λ. In this regime, the more appropriate ballistic theory also predicts a saturation of current for $L << \lambda$ (called the ballistic limit), but at a higher level and for different reasons[7][8]. In addition, present devices are more likely to operate in the transition regime where $L \sim \lambda$, referred to as the quasi-ballistic regime[7][8]. The ratio between the quasi-ballistic current and the drift diffusion current is named the Ballistic Enhancement Factor (BEF), a quantity of great

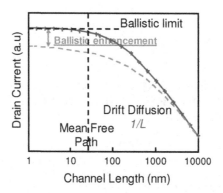

Figure 1: Schematic representation of the expected evolution of a MOSFET device on state current I_{on} versus channel length.

interest for device technology, always looking for any possible source of on state current enhancement.

The accurate evaluation of the BEF versus device characteristic requires highly sophisticated numerical models, accounting for quantum confinement within the channel and non-equilibrium transport physics, including all relevant scattering mechanisms. Extensive researches have been carried out in the last ten years to design such codes. To this purpose, two main physical models have been investigated. The first one consists in solving the Boltzmann Transport Equation, either by the Multi Subband Monte Carlo method[9][10][11] or by direct solving techniques[12]. In this semiclassical approach, the implementation of scattering mechanism is relatively well known and can be calibrated on experiments performed on large devices. However, longitudinal quantum effects can only be accounted for by means of subtle approximations. The second approach consists in solving the Schrodinger equation by the Non Equilibrium Green Function formalism[13][14], which rigorously captures the wave nature of electron and hole transport, but makes difficult the implementation of scattering mechanisms, especially when devices larger than few nanometers are considered. Despite huge efforts in the last years, these models are still in progress, especially to account for full band and mechanical strain effects. In addition, these codes are extremely time consuming, requiring extensive parallel computing, and are not available yet in commercial tools.

In this context, it is of great importance to develop approximated analytical models, which can capture the main features of quasi-ballistic transport. Such models could estimate in first order approximation what could be the Ballistic Enhancement Factor versus technological options. In addition, parameter extraction procedures from electrical measurements have also to be improved, in order to quantify the degree of ballisticity really achieved in advanced technologies. Both topics are addressed in this chapter.

The conventional Natori Lundstrom model of Quasi Ballistic transport will be described in the next section. Its limits are then investigated in paragraph 3. Finally, the experimental procedures used to quantify the degree of ballisticity in linear region will be discussed in the final section.

2. The Natori - Lundstrom models of Quasi Ballistic Transport

2.1. *The Natori model of ballistic transport*

Well known in the area of basic Physics[15], since the pioneering work of Landauer[16], the concept of ballistic limit has been re-investigated in the context of MOSFET devices by Natori[17] in 1994. His approach relies on the idea that transport into a ballistic device is no longer limited by the channel, but by the mechanism of carrier injection into it [8 15 17]. It is based on two main assumptions: 1/ device source and drain are supposed to be ideal reservoirs of carriers in equilibrium conditions, 2 / the gate is supposed to control perfectly the barrier between source and channel, as in well-designed device with negligible short channel effects. Under these hypothesis, the semi classical flux of carrier F_s^+ emitted from the source in equilibrium and entering through the channel (at a point called "virtual source") can easily be computed in a (100) Si electron channel, leading to:

$$F_s^+ = \frac{(2kT)^{3/2}}{\pi^2 \hbar^2} \left[\sum_i \sqrt{m_{cL}} F_{1/2} \left(\frac{E_{Fs} - E_i^L}{kT} \right) + \sum_i \sqrt{m_{cT}} F_{1/2} \left(\frac{E_{Fs} - E_i^T}{kT} \right) \right] \quad (1)$$

where $m_{cL} = m_t$, $m_{cT} = (m_l^{1/2} + m_t^{1/2})^2$. $F_{1/2}$ is a Fermi integral of order $\frac{1}{2}$, E_i^L (resp. E_i^T) are the unprimed (resp. primed) subband energies, i the subband index. In this one dimensional approach, in full ballistic regime, as the positive k states of the conduction band are populated by carriers emitted by the source, the carrier density N_s^+ flowing from source to drain is given by:

$$N_s^+(E_{Fs}) = \sum_i \frac{1}{2} \frac{m_{dL}}{\pi \hbar^2} kT \, F_0 \left(\frac{E_{Fs} - E_i^L}{kT} \right) + \sum_i \frac{1}{2} \frac{m_{dT}}{\pi \hbar^2} kT \, F_0 \left(\frac{E_{Fs} - E_i^T}{kT} \right) \quad (2)$$

where $m_{dL} = 2m_t$ and $m_{dT} = (m_l m_t)^{1/2}$. Similarly, as a difference of potential V_{ds} is applied between source and drain, the negative carriers density emitted by the drain and reaching the source end is given by

$$N_d^-(E_{Fs}, V_{ds}) = \sum_i \frac{1}{2} \frac{m_{dL}}{\pi \hbar^2} \frac{kT}{F_0} \left(\frac{E_{Fs} - qV_{ds} - E_i^L}{kT} \right) + \frac{1}{2} \frac{m_{dT} kT}{\pi \hbar^2} F_0 \left(\frac{E_{Fs} - qV_{ds} - E_i^T}{kT} \right) \quad (3)$$

In a well-designed MOSFET with negligible short channel effect, the charge at the virtual source remains constant when a bias V_{ds} is applied between source and drain. In consequence, the parameter E_{Fs} is adjusted in order to maintain a constant total charge Q_i at this point, as explained in details in references 18 and 19, solving the equation:

$$Q_i = q \, N_s^+(E_{Fs}) + q \, N_d^-(E_{Fs}, V_{ds} = 0) = q \, N_s^+(E_{Fs}) + q \, N_d^-(E_{Fs}, V_{ds}) \quad (4)$$

This procedure emulates the action of source – channel barrier modulation induced by the gate electrostatics. At $V_{ds} = 0$, E_{Fs} coincides with the inversion layer Fermi level. At last, the ballistic current I_d^{BAL} flowing from source to drain is simply given by:

$$I_d^{BAL} = q \, (F_s^+ - F_d^-) \tag{5}$$

where F_d^- is the flux of carrier emitted from the drain to the source. F_d^- has a similar expression than Eq. (1), except that, as carriers are emitted by the drain, the parameter E_{Fd} is equal to $E_{Fs} - qV_{ds}$.

Initially derived by Natori in the quantum limit regime (one single subband, completely degenerated), this model has been generalized in the more generalized case of multi subband inversion layer[18], and for various materials, arbitrary oriented[24]. To compute the energy level E_i entering in Eq. (1), the numerical solution of the coupled Poisson and Schrodinger equations at the virtual source is required. However, it can also be achieved by suitable analytical models, such as the models derived for bulk[21], Fully Depleted SOI[22] and double gate transistors[23].

In the subthreshold regime, this model only accounts for ideal thermionic emission in a well-designed MOSFET. A detailed modelling of the potential barrier between source and drain is thus required to include also the impact of short channel effects, band to band tunnelling and source to drain tunnelling [24][25].

2.2. *Injection velocity and subband engineering*

The ratio between the flux of carriers emitted by the source and entering the channel, divided by the corresponding carrier density is usually called the injection velocity V_{inj}. ($V_{inj} = F_s^+/N_s^+$). The injection velocity, computed by the Natori model, has been found in good agreement with the injection velocity extracted from Multi Subband Monte Carlo simulations (see figure 2), when devices featuring negligible short channel effects are considered. Note that in the high field conditions (corresponding to the transistor on state), as the drain is no longer emitting carriers capable of reaching the source, $I_d^{BAL} \sim qF_s^+ \sim Q_i V_{inj}$. In the ballistic regime, the Ballistic Enhancement Factor is thus simply given by:

$$BEF_{BAL} = \frac{v_{inj}}{v_{sat}} \tag{6}$$

where v_{sat} is the saturation velocity.

In weak inversion regime, the distribution of carrier at the virtual source follows a Maxwellian distribution. In this case, the injection velocity is equal to the thermal velocity, given for (100) silicon conduction band by:

$$v_{th} = \frac{2}{3}\sqrt{\frac{2\,kT}{\pi\,m_t}} + \frac{1}{3}\sqrt{\frac{2\,kT}{\pi\,m_l}} \approx 10^5 \, m/s \tag{7}$$

As already pointed out in ref. 7, in the case of silicon, the value of the thermal velocity, by pure hazard, is very close to the one of the saturation velocity v_{sat}. The consequence of such fortuity will be discussed later on. Note that this is usually not the case in other semiconductor materials, such as Germanium for instance.

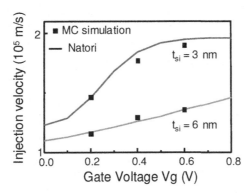

Figure 2: Comparison between injection velocity extracted from Multi Subband Monte Carlo simulations and calculated according the Natori model on undoped Double Gate MOSFET with silicon body t_{si}=3 nm (resp. 6 nm), channel length L=18 nm (resp. 28 nm), and t_{ox}=0.9 nm, V_d=0.6V (see ref. 19 for details).

In strong inversion regime however, the electron gas at the virtual source becomes degenerated. In this case, as high energetic states become more and more populated, the injection velocity tends to increase, as shown in Figure 2, exceeding the thermal velocity, and consequently the saturation velocity itself[19].

This phenomenon has received a considerable attention, as it is expected to increase the Ballistic Enhancement Factor. It is indeed possible, in principle, to further enhance the injection velocity by reducing the virtual source density of states (DOS), a procedure sometimes referred to as "subband engineering"[26]. Indeed, for the same amount of charges, states of higher energy would be more populated in a lower DOS than in a larger DOS device[19].

Several strategies are possible to reduce channel DOS. The first one would consist in reducing the number of populated subbands at the virtual source, by enhancing confinement. As seen in figure 3, the average injection velocity is indeed penalized by the contribution of other subbands, especially when they are not degenerated. Extremely thin SOI substrate can thus be used in order to reduce the number of populated subband[19,26].

Another technique consists in introducing mechanical strain[21,26], or simply using of low DOS alternative channel material [20,26,27,28,29]. Although the last option would be certainly the most effective in term of improvement of injection velocity (see figures 4 & 5), it would also require a radical change of the technology. This option is nevertheless currently extensively investigated at the research level[30,31].

Among the other "more conventional" options, the strain appears to be the most effective (see figure 4): an ideal biaxial strain for electrons for instance would lead to a 40 % improvement[19], while the enhancement of quantum effect due to the scaling of the body thickness down to 6 nm in Ultra Thin Body technologies would only lead to a 15 % improvement at best. The little impact of body thickness reduction is partially due to the effect of the wave function penetration through the gate dielectric due to tunneling[32,33], which tends to relax quantum confinement.

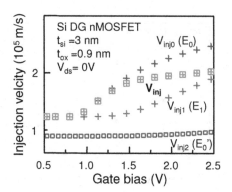

Figure 3: Gobal injection velocity versus gate voltage in double gate MOSFETs of 3 nm of body thickness. The injection velocity of the first three subbands is also shown for comparison, showing that the global injection velocity is lower than the first subband injection velocity, when the other subbands starts to be populated.

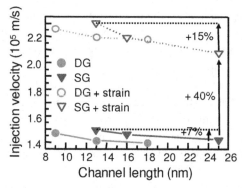

Figure 4: Injection velocity along the roadmap in Single Gate and Double gate MOSFETs devices, with and without biaxially strained channels

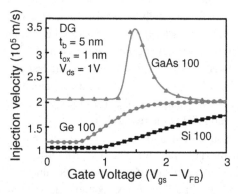

Figure 5: Injection velocity for nMOS Double Gate transistor, computed versus gate voltage, for (100) Si, (100) Ge and (100) GaAs materials. All relevant valleys Γ, Λ, Δ are included in the Poisson Schrodinger calculation

However, the implication of subband engineering investigated using the Natori model should been considered with care, for several reasons. First of all, the Ballistic

Enhancement Factor, which quantifies the enhancement of on state current due to ballistic transport versus drift diffusion simulation, is not the only figure of merit of a given technology. As far as CMOS is concerned, the I_{on}-I_{off} trade-off remains of course of primary importance. In this context, let us remind that first of all, low DOS devices usually also suffer, from the same reasons, to an enhancement of Dark Space phenomenon, which tends to further degrade the gate-to-substrate coupling[29][34][35][36]. In addition, alternative channel materials are also penalized by an increase of off state currents (Band to Band and Source to Drain Tunneling), especially at gate length below 15 nm[24][25][37]. Moreover, when the gate does not perfectly control the charge at the virtual source, which is unfortunately often the case in real devices, the injection velocity tends to further increase. In addition to DIBL, the virtual source itself can also be heated by field. These phenomena have been observed in several Monte Carlo simulations[38 - 45], and are not completely understood yet. Finally, the assumption of full ballistic transport still remains quite unrealistic[38 - 45]. For instance, results in ref. 40 have revealed that even a defect - free 10 nm undoped silicon channel cannot be considered as purely ballistic, and that the on state current has been indeed found 20 % lower than the ballistic current. Improvements of the Natori model to account for scattering will be thus discussed in the next section.

2.3. *Lundstrom models of backscattering*

To account for scattering, the Natori model has been improved by Lundstrom and co worker[7][8] using the "flux theory of transport", an approach initially introduced by McKelvey[46][47]. The key parameter of this new approach is the backscattering coefficient r, namely the ratio between the flux of carrier re injected to the source by scattering, divided by the flux of carrier injected by the source. This parameter can easily be introduced into the Natori model. First of all, assuming that it has the same value at the source and drain ends, the current flowing through the device can be expressed as:

$$I_d^{QBAL} = q \left[F_s^+ - r\, F_s^+ - (1-r)\, F_d^- \right] \tag{8}$$

The procedure for determining E_{Fs} has also to be modified, in order to account for backscattered carriers. In consequence, equation (4) becomes:

$$Q_i = q\, N_s^+ (E_{Fs})\, (1 + r(V_{ds})) + q\, N_d^- (E_{Fs}, V_{ds})\, (1 - r(V_{ds})) \tag{9}$$

Under particular bias conditions, these two equations can be further simplified. In ohmic regime, $Q_i \sim 2\, q\, N_s^+$. Assuming non degenerated statistics, and recalling that $V_{inj} = F_s^+/N_s^+ = v_{th}$, equation (8) simply reduces to:

$$I_d^{QBAL}{}_{lin} = (1 - r_{lin})\, \frac{Q_i}{2}\, v_{th}\, \frac{q\, V_{ds}}{kT} \tag{10}$$

In high field (saturation) regime however, the contribution to the total current of electron emitted by the drain can be neglected. In consequence:

$$I_d^{QBAL}{}_{sat} = \frac{1-r_{sat}}{1+r_{sat}}\, Q_i\, V_{inj} \tag{11}$$

The ballistic enhancement ratio, in the quasi-ballistic regime, is thus equal to:

$$BEF_{QBAL} = \frac{1-r_{sat}}{1+r_{sat}}\frac{V_{inj}}{V_{sat}} \tag{12}$$

Simple models have been proposed to estimate this backscattering coefficient. In low source to drain field condition, assuming a constant mean free path λ (average distance between two scattering events) and non degenerated statistics, it can be demonstrated using the flux theory[2][15] that:

$$r_{LF} = \frac{L}{L+\lambda} \tag{13}$$

In high field conditions, arguing that after a critical distance L_{kT}, scattering events would no longer be efficient enough to re-inject carrier back to the source because of the source to drain electric field attraction, the previous formula has been extended to high field condition, by substituting the channel length L by the critical distance L_{kT}, leading to:

$$r_{HF} = \frac{l_{kT}}{L_{kT}+\lambda} \tag{14}$$

This critical distance has been estimated as the distance needed by the potential to drop of a quantity of kT/q from the virtual source. Finally, the constant mean free path λ has been taken equal to:

$$\lambda = \frac{2\,\mu}{v_{th}}\frac{kT}{q} \tag{15}$$

where μ is the low field long channel mobility, a particular value that allow to match, both in high field and low field conditions, the ballistic (when L or $L_{kT} \ll \lambda$) and drift diffusion (when L or $L_{kT} \gg \lambda$) limit expressions (see figure 6). Formula (15) indicates that the low field long channel mobility μ is still a relevant parameter to improve performances, even in far from equilibrium regime of transport and in high field conditions. In addition, combining equations (10) (13) and (15) leads to a simple expression of the quasi ballistic current in the linear regime:

$$I_d^{QBAL}{}_{lin} = \frac{\mu'(L)}{L} Q_i V_{ds} \quad (16) \quad \text{with} \quad \mu'(L) = \mu\frac{L}{L+\lambda} \tag{17}$$

This result indicates that the apparent mobility μ' should be gate length dependent in quasi ballistic device, according to (17). This dependency has been confirmed by Monte Carlo simulations[38]. Note that this apparent mobility corresponds to the mobility extracted from experiments, using the usual Drift Diffusion formula.

Finally, let us note that the Lundstrom model, as recognized by the author himself[63], cannot be considered as a complete model. Indeed, the evaluation of the kT layer length requires knowing *a priori* the potential profile, which cannot be computed within this approach.

When drift diffusion equation applies, in long channel device, the potential profile can be analytically derived using the channel gradual approximation, leading to the following expression of the kT layer length:

$$L_{kT} = L\frac{2\ kT}{e(V_{dsat})} = L\frac{2\ kT}{e(V_g - V_T)} \tag{18}$$

As shown in figure 6, this equation is indeed the expression needed to match exactly the well-known long channel drift diffusion equation and the quasi-ballistic equation (11) in saturation regime. In shorter device, for compact model application, the kT layer is usually estimated using empirical formula calibrated on simulation results. In reference 64 for instance, the following expression has been used:

$$L_{kT} \approx L\left(\beta\frac{kT}{e(V_{dsat})}\right)^{\alpha} \tag{19}$$

where α is a parameter deduced from the exact shape of the potential profile ($\alpha \sim 0.7$). The β parameter takes into account degeneracy effects, and "should be somewhat greater than 1"[64].

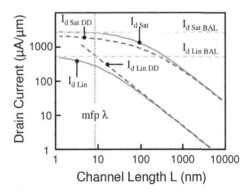

Figure 6: Id – L curves (at same voltage, and constant mean free path) computed in linear ($V_{ds} = 10$ mV) as well in saturation regime ($V_{ds} = 1$ V) using the Natori Lundstrom approach (Id_{Lin} and Id_{Sat}) and the drift diffusion approach ($Id_{Lin\ DD}$ and $Id_{Sat\ DD}$), accounting for saturation velocity. The ballistic limit ($Id_{Lin\ BAL}$ and $Id_{Sat\ BAL}$) is also shown for comparison. ($\mu = 200$ cm^{2}V^{-1}s^{-1}, $V_{inj} = 1.2\ 10^5$ m/s, $N_{inv} = 1.45\ 10^{13}$ cm^{-2}, L_{kT} has been estimated using equation (18)).

The Natori Lundstrom formalism, with or without minor improvements, has become extremely popular in the last ten years. It has been considered as the state of the art of the understanding of transport in advanced MOSFETs, to analyze experiments[48 49 50 51 52 53 54]

[55] [56] [57], improve compact models[58] [59] [60] [64] as well as to investigate scaling trends[61] [62]. The validity and limits of this approach are discussed in the next section.

3. Beyond the Natori-Lundstrom model

3.1. *Theoretical foundations of the Natori Lundstrom model: the quasi ballistic drift-diffusion theory*

Due to the empirical nature of the arguments introducing it and despite quantitative validation by Monte Carlo simulations[39] [42], the concept of kT layer has raised skepticism among the device modeling community. In order to achieve a better understanding of the kT layer concept, its theoretical basis has been investigated in reference 65. Considering one dimension in space, assuming non degenerated statistics, one single isotropic band, and treating the collision integral by a simple "relaxation length" approximations, its has been shown that the Boltzmann Transport Equation can be reduced to a simpler formalism, sometimes referred to as the "Quasi Ballistic drift diffusion" formalism. Within these approximations, the solution of the BTE has shown that the distribution function f(x,v) could be artificially split into two functions, one for positive velocity, one for negative velocity, both of them having a Maxwellian (thermal) shape, given by:

$$f^{\pm}(x, v_x) = 2\sqrt{\frac{m}{2 \pi kT}} \; n^{\pm}(x) \; \exp\left(-\frac{m v_x^2}{2 kT}\right) \tag{20}$$

where n^+ (resp. n^-) are the concentration of carriers flowing from source to drain (resp. from drain to source). In addition, these concentrations have been found to obey the following conservation equations:

$$\frac{d\,n^-}{dx} + \frac{n^-}{kT}\frac{d\,U}{dx} = \frac{n^- - n^+}{\lambda} \tag{21}$$

$$\Phi^+(x) - \Phi^-(x) = \left(n^+(x) - n^-(x)\right)v_{th} = \Phi \tag{22}$$

where Φ is the net flux of carriers, independent of the position x. This approach, approximation of semi classical transport in the framework of the relaxation length approximation, includes the impact of both isotropic scattering and arbitrary field, and constitutes thus a more general formalism than the kT layer model. It has been known in fact for a long time[66] [67], but its connection with the kT layer concept has been clarified only recently[65].

Indeed, an expression of the backscattering coefficient can be derived using the quasi ballistic drift diffusion model, assuming a linear potential profile $V(x) = -q\,F\,x$ (with F < 0) between source and drain. In order to discriminate backscattered carriers emitted by the source from carriers emitted from the drain, the drain reservoir has been assumed in the calculation to only absorb carriers, leading to the boundary condition $n^-(L) = 0$. As the calculations are performed non-self consistently, this procedure does not impact the expression of r, as demonstrated in 65.

Solving equations (21) and (22) leads to the following expression of the backscattering coefficient r:

$$r = \frac{j^-(0)}{j^+(0)} = \frac{v_{th} \; n^-(0)}{v_{th} \; n^+(0)} = \frac{L_{kT} \; (1-\beta)}{\lambda + L_{kT}(1-\beta)} \tag{23}$$

with $\beta = \exp(-L / L_{kT})$, and $L_{kT} = kT / q \mid F \mid$. This simple equation of r nicely tends to the Lundstrom formula at both low field (Eq. 13) and high field (Eq. 14) conditions.

Let us examine in more details the implications of this "quasi ballistic drift diffusion model". First of all, starting from equation (21), multiplying by v_{th}, and introducing the total carrier concentration n, equal to:

$$n(x) = n^+(x) + n^-(x) \tag{24}$$

the net flux of carriers Φ can be re-written as :

$$\Phi = -D\frac{d\,n}{dx} - 2\,n^-\,\mu E \tag{25}$$

where $D = \lambda \, v_{th}$ and $\mu = e \, D / kT$. Using equation (15) and Einstein relation, μ and D effectively correspond to the conventional long channel low field mobility and diffusion coefficient. Moreover, n^- in equation (25) can be expressed as a function of n and Φ, using both equations (22) and (24), leading to:

$$\Phi = -D'\frac{d\,n}{dx} - n\,\mu' E \tag{26}$$

where

$$\mu' = \frac{\mu}{1 + \mu E / v_{th}} \tag{27}$$

and $\mu' = e \, D' / kT$. Equations (26) and (27) suggest that one of the consequences of the quasi ballistic drift diffusion model, compared to the conventional drift diffusion approach, is the introduction of a longitudinal field dependent mobility. In addition, equation (27) limits the average velocity to the maximum value of v_{th}. In this approach, this modification is a direct consequence of the constant mean free path and relaxation length approximations, which force the positive and negative distribution functions to keep a Maxwellian shape along x. Thus positive and negative carriers move with an average velocity equal to the thermal velocity. Ironically, conventional drift diffusion already includes longitudinal field dependency to the mobility, in an attempt to account for saturation velocity. As already pointed out by Lundstrom et al. [7 8], as in Silicon the thermal velocity and the saturation velocity have similar values, it explains *a posteriori* why a simple drift diffusion model can qualitatively emulate the ballistic limit.

However, it should be mentioned that boundary conditions have to be applied with care when using the quasi ballistic drift diffusion model. Indeed, as discussed in 65, and contrary to what has been done in previous works such as 60, in the quasi ballistic model, boundary conditions are applied on $n^+(0)$ and $n^-(0)$ and not on $n(0)$ and $n(L)$, as in a conventional drift diffusion model. These quasi ballistic boundaries conditions make non obvious the direct use of equations (26) and (27). Note that if conventional drift diffusion boundary conditions were used instead of the correct one, the impact of quasi-ballistic transport particularly in low field condition would be erroneous. Indeed, as $\mu E \ll v_{th}$ in low field condition, this approach would lead to the conclusion that the apparent mobility in the quasi ballistic regime remains equal to the long channel mobility μ, even in a full ballistic channel. This is obviously not the correct result, given instead by equation (17).

Finally, let us note that equation (27) constitutes one of the main limitations of the quasi ballistic drift diffusion model, as it suppresses any possibility of velocity overshoot. Indeed, as it will be more clearly shown in the next section, velocity profiles obtained by Monte Carlo simulations usually largely exceed the thermal velocity. As the potential profile should be computed self consistently with the motion of carriers (thus the velocity profiles), it induces that the potential profile, and thus the L_{kT} value computed self consistently, is indeed erroneous in the quasi ballistic drift diffusion approach.

The quasi ballistic drift diffusion model approach[65][66][67] can take different forms in the literature. As already mentioned, it is equivalent to the Lundstrom backscattering coefficient model[7][8]. The equivalence with the Gildenblat flux model [58][59] can also be proved.

3.2. *Comparison with Monte Carlo simulations: results and discussion*

The validity of the Lundstrom formula of backscattering have been investigated by Monte Carlo simulations in several contributions[38][39][40][42][43][44][45]. This section summarizes the conclusions obtained in one of the most recent works[42].

In this paper, mobility μ and backscattering coefficient r have been computed using the Monte Carlo (MC) method, in simplified structures. These template devices are 1D in real space, and 2D in the momentum space. The effective mass approximation has been used, assuming a spherical band (with an effective mass equal to m_0). Only phonon scattering has been taken into account, featuring one acoustic phonon mode and one optical phonon mode (of energy 35 meV). Simulations have been performed in a "frozen field" mode, i.e. without computing the potential energy profile self consistently with the motion of carriers. These simplifications have only been made in an attempt to simplify the analysis and interpretation of results. The mobility has been computed at low field condition in an infinitely long structure, by imposing periodic boundary conditions. In addition, backscattering coefficient has been simulated in a finite structure of length L, where the right contact (drain-like) has been artificially "switched off", i.e. does not inject carriers into the structure, as explained in the last section. In term of MC simulation, it means that particles are only injected by the left contact (source like), assumed to be in equilibrium condition, and absorbed by the drain. This unphysical boundary condition makes easier the extraction of the backscattering coefficient, especially at low field condition, as the flux of carriers coming back to the source have been necessarily emitted by the same contact (and not by the drain). While the Potential

Energy profile has been taken equal to zero at low field condition, both a linear and parabolic Potential Energy profiles have been considered at high field condition.

The backscattering coefficient r extracted at low field condition has been plotted versus the structure length L in figure 7. It turns out that the MC results can be nicely fitted by equation r = L / (L+λ) (dotted lines in figure 7), provided that the mean free path λ has been used as a fitting parameter. The low field mean free path extracted by this procedure will be referred to as λ_0 in the following.

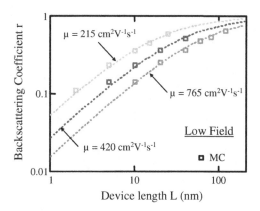

Figure 7: Low Field backscattering coefficient r_{LF} versus device length L, extracted from 1D non self consistent Monte Carlo simulation in low field condition, featuring different acoustic phonon coupling constants, i.e. different mobilities. The coupling constant for optical phonons has been kept constant and equal to 2.10^{12} eV/m.

A similar procedure has been applied to the high field results: r has been plotted as a function of the kT layer length L_{kT}, as shown in figure 8. Again, these results can be accurately reproduced by r = L_{kT} / (L_{kT} +λ) using the mean free path (referred to as λ_F in the following) as a fitting parameter.

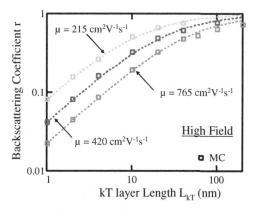

Figure 8: High Field backscattering coefficient r_{HF} versus kT layer length L_{kT}. The device length is L = 10 L_{kT}. The L_{kT} range corresponds to electric fields from 2.6kV/cm to 260kV/cm. Other parameters are the same than in Figure 5.

Both λ_0 and λ_F have been plotted in figure 9, as a function of the mobility also computed by MC simulations in the same but infinitely long simple structure. According to the Lundstrom theory, λ_0 and λ_F should be equal and given by equation (15), also plotted for comparison in dotted line in figure 9. While λ_0 has been found in qualitative good agreement with the prediction of equation (15), λ_F however appears significantly lower.

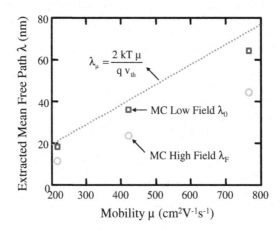

Figure 9: Extracted mean free path in low field (square) and high field (circle) conditions, plotted versus low field mobility. Equation (15) is also shown in dotted line as reference.

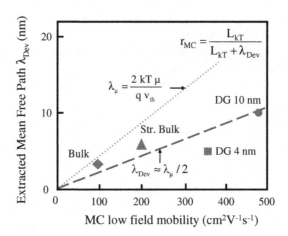

Figure 10: Extracted mean free path in high field condition extracted from device Monte Carlo simulations (symbol) plotted versus MC long channel low field mobility. Equation (3) is also shown in dotted line as reference.

Similar comparisons have also been performed on more realistic device structure (see figure 10) (2D in real space, including all the relevant scattering mechanisms: phonons, impurities, surface roughness and body thickness fluctuations in the case of ultra thin body devices). All devices are 25 nm long, "Bulk" refers to unstrained bulk transistor,

"Str. Bulk" to a similar device with ideal biaxially strained channel, DG to undopded double gate devices with body thickness of 10 nm and 4 nm. Device mobility has been computed using the same model, but in an infinitely long inversion layer. It turns out that even using a more sophisticated model, the extracted mean free path in high field condition appears qualitatively proportional to the corresponding long channel low field mobility, however with a lower slope than expected according to Eq. (15).

In conclusion, the Lundstrom model for backscattering coefficient may appear qualitatively correct, especially in low field condition, but also in high field, provided to use the mean free path as a fitting parameter. This extracted mean free path has been found shorter that the equilibrium mean free path predicted by the Lundstrom theory.

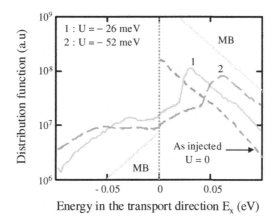

Figure 11: Distribution functions as injected (dotted line) and at different point of the template structure, close to the emitting source, as a function of the energy E_x in the transport direction. The thermal Maxwellian shapes are also indicated for comparison.

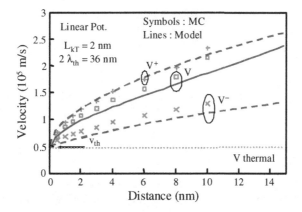

Figure 12: Positive V^+, negative V^- and average V Monte Carlo Velocity profile (symbols) versus distance in a template 1D structure with linear profile potential (in high field conditions) and absorbing drain. Results obtained by the analytical model proposed in 69 (dotted line) are shown for comparison. $V_{thermal}$ indicates the velocity profile obtained using the quasi ballistic drift diffusion model.

The detailed understanding of this discrepancy at high field is however more complex than it may seem. This point has been investigated in more detailed in ref. 69, underlying the role of heated distribution functions on quasi ballistic transport. Indeed, as explained in section 3.1, the kT layer approach approximates the carrier distribution function f by two equilibrium distribution functions (one for positive, one for negative velocity), following equations (20). As shown in figure 11, this assumption is of course in disagreement with Monte Carlo simulations, which however predicts a distribution function closer to a pure ballistic one. This limitation of the kT layer approach is particularly penalizing to model the velocity profiles. Indeed, according the thermal distribution functions given by equations (20), the carrier velocity can never exceed the thermal velocity. This is obviously not true, in particular in the high field region close to the drain (see an example Fig. 12).

In 69, the balance equations of transport in the relaxation length approximation has been generalized to any kind of distributions functions, and solved for approximated distribution functions inspired by Monte Carlo results. This approach has lead to backscattering coefficient in much better agreement with simulations, without any artificial reduction of the mean free path. However, efforts are still needed to turn this kind of approaches into a complete compact model.

4. Electrical Characterization of MOSFETs in the Quasi Ballistic Regime

4.1. *Introduction & State of the art*

The analytical modeling of quasi ballistic transport, reviewed in the previous sections, has raised of course several interrogations from an experimental point of view:

- first of all, is there an experimental way to validate or not the concepts of Ballistic, Quasi Ballistic Transport and formula for backscattering coefficients?
- assuming that the quasi ballistic theory applies, how to improve parameter extraction procedures, which usually rely on the "old fashion" drift diffusion equations?
- and more specifically how to monitor the Ballistic Enhancement Ratio, possible "booster" of CMOS performances?

In previous works, the quasi ballistic theory of transport has been implicitly assumed as valid, and most of the attention has been focused on the two last questions. In particular, several works have tried to define a suitable parameter extraction procedure to measure the backscattering coefficient r or the ballistic ratio, usually at high field conditions[48 49 50 51 52 53]. In addition to the usual experimental difficulties (series resistance extractions, capacitance measurements …), most of these techniques have required an *a priori* knowledge of either the ballistic limit or the injection velocity. As mentioned before, the available analytical approaches to estimate them are not very accurate, leading, as explained in ref. 38, to significant errors in parameter extractions. In addition, it was never easy to benchmark results obtained in different works, as the extracted data significantly depends on the model used to estimate the ballistic limit or the injection velocity.

Moreover, other works have tried to determine some experimental evidence of the quasi ballistic nature of transport in advanced MOSFETs [54] [55] [56] [57]. These works have raised several relevant doubts and questioning, not about the theory itself, but rather about the applicability of the theory to advanced MOSFETs. The methodology and results of one of these works are reported in the following section.

4.2. *Principle of backscattering coefficient extraction in the linear regime*

Following reference 72 and equations (16) and (17), the quasi-ballistic drain current of a MOS transistor in linear operation can readily be equated to[73]:

$$I_d = (1-r)\frac{W}{L}\mu_{bal}Q_iV_d = r\frac{W}{L}\mu_{dd}Q_iV_d = \frac{W}{L}\mu_{exp}Q_iV_d \qquad (28)$$

where Q_i is the inversion charge, W and L the gate width and length, V_d the drain voltage, μ_{bal} the "ballistic" mobility, μ_{dd} the drift-diffusion one (i.e. the low field long channel mobility) whereas μ_{exp} stands for the experimental or apparent mobility to be measured from drain current, by using Eq. 28. The ballistic mobility μ_{bal} can be derived from equation (28) after considering the drain current expression in the ballistic limit equations (1)-(5), considering only one subband, yielding [74] [75],

$$\mu_{bal} = \frac{e\,v_{th}}{2kT}\cdot\frac{L}{F_0(\eta_F)}\frac{F_{-1/2}(\eta_F)}{F_0(\eta_F)} \qquad (29)$$

where v_{th} is the thermal velocity at the virtual source, $F_{-1/2}$ and F_0 are the Fermi-Dirac functions, with η_F being the reduced Fermi level $(E_0 - E_F)/kT$. In the Boltzmann statistics limit, the ratio $F_{-1/2}(\eta_F) / F_0(\eta_F)$, also called the degeneracy factor $DF(\eta_F)$, reduces to one, recovering Shur's μ_{bal} expression[74].

Note that the degeneracy factor can be well approximated as $DF(Q_i) \approx 1 / (1 + Q_i/Q_c)$ for the single subband case in the quantum limit (Q_c being a constant close to 2.10^{13} q/cm² for silicon). Figure 13 shows a simulation of the DF function for (100) silicon when

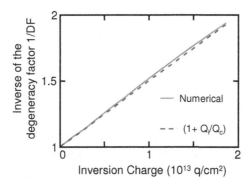

Figure 13: Variation of 1/DF with inversion charge Q_i for the fundamental 2D subband of (100) silicon : exact (solid line) and linear approximation (dashed line). Parameters: gate oxide thickness t_{ox} = 1.6nm, Q_c=2.10^{13}q/cm².

considering only the fundamental subband, which validates the previous empirical relationship.

Eliminating the backscattering coefficient in Eq. (28) enables to recover the Matthiessen-rule-like expression for μ_{exp}[74]:

$$\frac{1}{\mu_{exp}} = \frac{1}{\mu_{bal}} + \frac{1}{\mu_{dd}} \qquad (30)$$

It is now straightforward to derive the backscattering coefficient from Eq. (28) as :

$$r = 1 - \frac{\mu_{exp}}{\mu_{bal}} \qquad (31)$$

4.3. Results and discussion

The above method has been used for the extraction of the backscattering coefficient and drift-diffusion mobility in advanced CMOS devices fabricated by STMicroelectronics (Crolles). Bulk and FD-SOI nMOS devices were tested. The bulk devices are representative of a 65nm CMOS technology featuring a doped channel ($\approx 10^{17}/cm^3$) with halos, SiON gate oxide (CET=2.2nm) and polysilicon gate. For the FD-SOI technology, the nMOS devices were made on 300 mm <100> UNIBOND[TM] SOI wafers with a BOX of 145 nm. The SOI films were thinned down by thermal oxidation and wet etched to achieve a final thickness around 10 nm under the gate at the end of the process. After STI isolation, a HfSiON dielectric of approximately 2.5 nm was deposited. A TiN of 10nm and a poly-Si layer of 100 nm were deposited for gate fabrication. A 193 nm lithography combined with trimming was performed to achieve reduced gate dimensions. The minimum gate length dimension measured on the wafers is around 40nm. After an offset spacer of 10nm realization, a selective epitaxy of 10nm was performed in order to reduce access resistance and to facilitate NiSi S/D salicidation. Raised extensions were implanted. To finish, a Dshape spacer, S/D implantation and salicidation were realized. The device channel was left undoped.

The static parameter extraction was then performed in linear operation regime (V_d = 20mV) on transistor arrays with common source and gate for various gate lengths (40 nm to 10 μm) using the $Y = I_d/g_m^{1/2}$ function method[76], allowing the elimination of series resistance effect, the extraction of threshold voltage and of the low field mobility μ_0 for each gate length. The effective gate length and the gate oxide capacitance C_{ox} were extracted from gate-to-channel capacitance measurements[77].

The low field mobility μ_{exp} was extracted as a function of channel length for the various tested nMOS devices (see Figure 14). Note that in all cases, a strong degradation of the mobility, by about a factor 2, is observed as the channel length is reduced below 100nm for both technologies. The drift-diffusion contribution of the mobility μ_{dd} was then evaluated after subtracting the ballistic mobility contribution using Eq. 30. The ballistic mobility μ_{bal} has been calculated using Eq. 29, assuming that the injection velocity is equal to the thermal velocity. This is a reasonable assumption since the low field mobility is experimentally extracted near threshold voltage, i.e. at low enough inversion charge. Figure 14 demonstrates without any ambiguity that the drift-diffusion mobility μ_{dd} is strongly degraded below 100nm, and that the ballistic effects cannot explain here the huge apparent mobility reduction obtained on both technologies. This mobility behavior is a general feature of both bulk and thin film devices regardless of their channel doping as was already observed in gate-all-around (GAA) and bulk MOS structures[54 78 79]. This could be interpreted as an increasing contribution of scattering mechanisms in shorter devices (below 70 - 80nm), possibly due to neutral defects. These defects may originate from source-drain implantation-induced Si interstitials in undoped film and halo extra doping in bulk architectures[57].

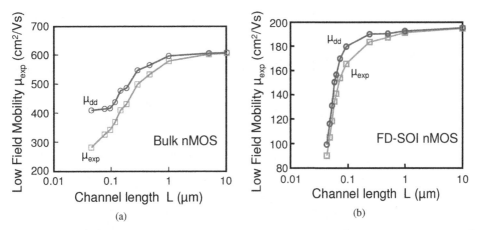

Figure 14: Variation of experimental low field mobility μ_{exp} with channel length L as obtained on bulk (a) and FD-SOI devices (b). The drift-diffusion mobility μ_{dd} obtained after ballistic effect correction is also shown.

Finally, the backscattering coefficient r has been extracted using Eq. 31 as a function of gate length (Figure 14). As expected, r is decreasing at small gate length down to about 0.70 and 0.86, which indicates that the ballistic ratio in ohmic regime (i.e $1 - r$) in these devices reaches at most 30% and 14% over 40-50nm gate length range, for bulk and FD-SOI devices, respectively.

The larger ballistic rate observed in the bulk devices can be justified by the higher mobility values over the whole gate length range, despite the channel and halo higher doping levels. This feature clearly indicates that the quality of undoped thin film structures is not yet optimized compared to well mature bulk technologies in order to benefit from full ballistic effect.

The different behavior of μ_{dd} and r for bulk and FD-SOI devices at small channel length, could be attributed to the fact that, for bulk devices, halos are merging below 0.1μm, yielding a nearly constant channel doping. In consequences, μ_{dd} in bulk devices is becoming constant at small gate length. In contrast, for FD-SOI devices, μ_{dd} is still degraded at small lengths because the concentration of the S-D implantation-induced defects should increase as getting closer to source and drain junctions.

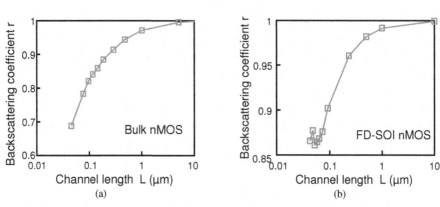

Figure 15: Variation of backscattering coefficient r with channel length L for Bulk (a) and FD-SOI devices (b).

This mobility degradation as a function of gate length has recently been analyzed in a statistical way and it has been found, as illustrated in Fig. 16 for FD-SOI devices, that the low field mobility can always be expressed as[79],

$$\frac{1}{\mu_0} = \frac{1}{\mu_{dd}} + \frac{\alpha_\mu}{L} \tag{32}$$

where μ_{dd} is the low field long channel mobility and α_μ a mobility degradation factor [nm.V.s/cm²]. Actually, Eq. 32 has successfully been applied to a large panel of CMOS devices featuring bulk, FD-SOI, double gate or GAA architectures[79]. Interestingly, as can be seen from Fig. 17, a strong correlation exists between the maximum mobility (for long devices) and the degradation factor α_μ. The comparison of the experimental results given by equation (32) with Eqs 29 and 30 reveals that α_μ has a minimum theoretical value,

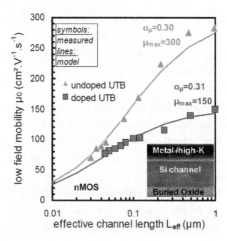

Figure 16: Comparison of electron mobility between undoped and doped ultra thin body (UTB) with high-K/metal gate stack (after Bidal et al[79]).

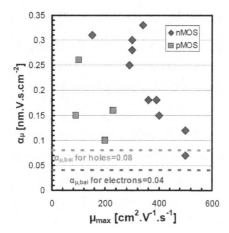

Figure 17: α_μ as a function of μ_{max} clearly showing no universal correlation between α_μ and μ_{max} (after Bidal et al[79]).

given by its ballistic limit $(2kT/q)/v_{th}$. (for Boltzmann's statistics). The data of Fig. 17 also indicate that this limit could only be reached on few devices, which feature a very high long channel mobility of 600 cm^2/Vs (resp. 300 cm^2/Vs) for electrons (resp. holes).

5. Conclusions

Since the pioneering works of K. Natori[17] in 1994, and then M. Lundstrom[7] in 1997, the quasi ballistic regime of transport has become an extremely popular field of research in the area of MOSFET device physics. Interestingly, these researches have mostly been focused on the understanding of it, by means of analytical modeling and electrical characterization, since it was already naturally included in numerical sophisticated tools such as Boltzmann Transport Equation or Schrodinger Equation solvers.

Despite several ongoing controversies and unsolved issues [44] [45], the main conclusions of these works have already become a new way of understanding performance enhancements of nano MOSFETs. For instance, the growing interest for high mobility channel devices[30] [31] is usually presented in terms of "mean free path" and "injection velocity" enhancements, two typical quasi ballistic concepts.

The quasi ballistic analytical models have been thus reviewed in this chapter.

First of all, the analytical modeling of MOSFET ballistic limit, following the approach of Natori has been presented. These theories have been used to investigate the "subband engineering[26]", which consists in enhancing ballistic limit and thus improving injection efficiency, by raising 2D carrier gas confinement (introducing strain, film or field confinement, or new channels materials).

The critical impact of scattering on performances has also been discussed, presenting the "orthodox" Lundstrom's approach[8], and its limitations. In particular, the connections between the high field kT layer backscattering theory and the old fashion "saturation velocity" have been clarified.

Despite progresses in analyzing the success and limitations of the backscattering theories, efforts are still to be made to achieve a simple formalism, able to fulfill the requirements of circuit oriented compact model, and accounting for ballistic limit, scatterings, velocity overshoot and self-consistency with electrostatic.

A closer look to experiments has also revealed an interesting feature of quasi ballistic transport: it is not clear, from an experimental perspective, if this phenomenon actually occurs or not. Indeed, measurements performed on several different technologies suggest a degradation of transport when reducing the gate length, making unlikely the existence of quasi ballistic regime. These degradations could possibly result from neutral defects[54] generated by source and drain implantations. Again, more investigations are needed to confirm the existence of such defects, which challenge our understanding of advanced MOSFETs devices.

Acknowledgments

This work has been partially supported by the Nanosil Network of Excellence. The authors would like to thank colleagues for there contributions to the research activities presented here: Pierpaolo Palestri and Luca Selmi from Udine University (Italy), Antoine Cros, Frédéric Boeuf, Stéphane Monfray and Thomas Skotnicki from STMicroelectronics Crolles, and PhD students : Marlene Ferrier, Luca Lucci, Quentin Rafhay, Illias Pappas, Dominique Fleury and Gregory Bidal.

References

1. ITRS 2003 Public Home Page, « Process Integration, Devices and Structures», http://www.itrs.net/
2. M. Lundstrom, "Fundamentals of Carrier Transport", 2nd Ed. Cambridge University Press, Cambridge UK, 2000
3. J. A. Cooper, D. F. Nelson, "High-field drift velocity of electrons at the Si–SiO$_2$ interface as determined by a time-of-flight technique" *J. Appl. Phys.* **54** (3), p. 1445–56 (1983)
4. C. Jacoboni, L. Reggiani, "The Monte-Carlo method for the solution of charge transport in semiconductors with applications to covalent materials" *Rev Mod Phys* **55** (3) p. 645–705 (1983).
5. D. M. Caughey and R. E. Thomas, "Carrier Mobilities in Silicon Empirically Related to Doping and Field," *Proc. IEEE* **55**, (12), p. 2192-2193 (1967).
6. S. E. Laux, R. G. Byrnes, "Semiconductor device simulation using generalized mobility models", *IBM J. Res. Dev.* **29**(3) p.289-301 (1985)
7. M. Lundstrom « Elementary Scattering theory of the Si MOSFET » *IEEE Electron Device Lett.* **18** p. 361-363 (1997).
8. M. Lundstrom, Z. Ren, « Essential Physics of Carrier Transport in Nanoscale MOSFETs » IEEE *Trans. on Electron Devices*, **49** p. 133 - 141 (2002).
9. M. Fischetti et S. Laux., "Monte Carlo study of electron transport in silicon inversion layers", *Phys. Rev.* B, p. 2244 (1993).
10. L. Lucci, P. Palestri, D. Esseni, L. Selmi, "Multi-subband Monte Carlo modeling of nano-MOSFETs with strong vertical quantization and electron gas degeneration," in *IEDM Tech. Dig.* pp.631, 2005

11. L. Lucci, P. Palestri, D. Esseni, L. Bergagnini, L. Selmi; "Multisubband Monte Carlo Study of Transport, Quantization, and Electron-Gas Degeneration in Ultrathin SOI n-MOSFETs", *IEEE Trans. Electron Devices*, **54**(5), pp. 1156–1164, May. 2007.

12. S.-M. Hong, C. Jungemann "A fully coupled scheme for a Boltzmann-Poisson equation solver based on a spherical harmonics expansion" *J. Comp. Elec.* **8**(3-4) p. 225-241 (2009)

13. S. Datta, "Nanoscale device modeling: The Green's function method", *Superlatt. Microstruct.*, **28**, pp. 253–278, (2000)

14. M. Luisier, A. Schenk, "Atomistic Simulation of Nanowire Transistors", *J. Comput. Theor. Nanosci.* **5**(6), p. 1031–1045 (2008)

15. S. Datta, « Electronic transport in mesoscopic Systems » Cambridge University Press, Cambridge UK, 1997.

16. R. Landauer, "Spatial Variation of Currents and Fields Due to Localized Scatterers in Metallic Conduction," *IBM J. Res. Dev.* **1**, p. 233 (1957).

17. K. Natori, "Ballistic metal-oxide-semiconductor field effect transistor", *J. Appl. Phys.*, **76** (8), p. 4879–4890, October 1994.

18. F Assad, Z Ren, D Vasileska, S Datta, M Lundstrom. "On the performance limits for Si MOSFETs: a theoretical study." *IEEE Trans. Electron Devices* **47**(1) p.282 (2000)

19. M. Ferrier, R. Clerc, L. Lucci, Q. Rafhay, G. Pananakakis, G. Ghibaudo, F. Boeuf, T. Skotnicki., « Conventional Technological Boosters for Injection Velocity in Ultra Thin Body MOSFETs » *IEEE Trans. on Nanotechnology*, **6**(6),p. 613 – 621 (2007)

20. Q. Rafhay, R. Clerc, M. Ferrier, G. Pananakakis and G. Ghibaudo, « Impact of Channel Orientation on Ballistic Current of nDGFETs with Alternative Channel Materials » *Solid-State Electronics*, **52**(4),p. 540-547 (2008)

21. M. Ferrier, R. Clerc, G. Ghibaudo, F. Boeuf, T. Skotnicki, "Analytical model for quantization on strained and unstrained bulk nMOSFET and its impact on quasi-ballistic current" *Solid-State Electronics*, **50** p. 69-77 (2006)

22. M. Ferrier, R. Clerc, L. Lucci, G. Ghibaudo, A. Vandooren, F. Boeuf, T. Skotnicki, « Saturation Drain Current analytical modeling of Single Gate Fully Depleted SOI or SON MOSFETs in the Quasi Ballistic Regime of Transport », *proc. of the IEEE International SOI Conference*, USA, Niagara Falls, 2 – 5 October, 2006

23. M. Ferrier, R. Clerc, G. Pananakakis, G. Ghibaudo, F. Bœuf, T. Skotnicki, « Analytical Compact Model for Quantization in Undoped Double-Gate Metal Oxide Semiconductor Field Effect Transistors and Its Impact on Quasi-Ballistic Current », *Jpn. J. Appl. Phys.* **45** Part 1, No. 4B p 3088-3096 (2006)

24. Q. Rafhay, R. Clerc, G. Ghibaudo, G. Pananakakis, « Impact of source-to-drain tunnelling on the scalability of arbitrary oriented alternative channel material nMOSFETs », *Solid-State Electronics*, **52**(10), p. 1474 - 1481 (2008)

25. Q. Rafhay, R. Clerc, G. Pananakakis and G. Ghibaudo, « Source-to-Drain vs. Band-to-Band Tunneling in Ultra-Scaled DG nMOSFETs with Alternative Channel Materials » Proc. SSDM 2008 pp. 36 - 37 (2008)

26. S. Takagi, "Re-examination of Subband Structure engineering in Ultra-Short Channel MOSFETs under Ballistic Carrier Transport" in *Proc. VLSI 2003*, pp. 115-116.

27. S. Takagi, T. Mizuno, T. Tezuka, N. Sugiyama, S. Nakaharai, T. Numata, J. Koga, K. Uchida, "Sub-band Structure engineering for Advanced CMOS channels", *Solid-State Electronics*, **49**, pp. 684-694, 2005

28. A. Pethe, T. Krishnamohan, D. Kim, S. Oh, H. S. P. Wong, Y. Nishi, K. C. Saraswat, "Investigation of the performance limits of III-V Double-Gate n-MOSFETs", in *IEDM Tech. Dig.* pp. 605 (2005)

29. M. De Michielis, D. Esseni, F. Driussi, "Analytical Models for the Insight Into the Use of Alternative Channel Materials in Ballistic nano-MOSFETs" *IEEE Trans. Electron Devices* **54**(1) p.115-123 (2007).

30. K. Saraswat, C. O. Chui, K. Donghyun, T. Krishnamohan, A. Pethe, "High mobility materials and novel device structures for high performance nanoscale MOSFETs," in *IEDM Tech. Dig.*, pp. 659-662 (2006)

31. G. Dewey, M. K. Hudait, K. Lee, R. Pillarisetty, W. Rachmady, M. Radosavljevic, T. Rakshit, R. Chau, "Carrier Transport in High-Mobility III–V Quantum-Well Transistors and Performance Impact for High-Speed Low-Power Logic Applications" *IEEE Electron Device Lett.*, **29**, p. 1094 (2008)

32. S. Mudanai, L. F. Register, A. F. Tasch, and S. K. Banerjee, "Understanding the Effects of Wave Function Penetration on the Inversion Layer Capacitance of nMOSFETs" *IEEE Electron Device Lett.* **22**, 145 (2001).

33. M. Yunus and A. Haque 'Wave function penetration effects on current–voltage characteristics of ballistic metal–oxide–semiconductor transistors', *J. Appl. Phys.* **93**, p. 600 (2003)

34. Q. Rafhay, R. Clerc, J. Coignus, G. Pananakakis, G. Ghibaudo, « Dark Space, Quantum Capacitance and Inversion Capacitance in Si, Ge, GaAs and $In_{0.53}Ga_{0.47}As$ nMOS Capacitors », proc. of the ULIS conference pp. 33-37 (2010)

35. H. S. Pal, K. D. Cantley, S. S. Ahmed, M. S. Lundstrom, "Influence of Bandstructure and Channel Structure on the Inversion Layer Capacitance of Silicon and GaAs MOSFETs" *IEEE Trans. Electron Devices*, **55**, (3), pp.904-908 (2008).

36. D. Jin, D. Kim, T. Kim, J. A. Del Alamo, "Quantum Capacitance in Scaled Down III-V FETs", in *IEDM Tech. Dig.* pp. 495-499, (2009).

37. K. D. Cantley, Y. Liu, H. S. Pal, T. Low, S. S. Ahmed, and M. S. Lundstrom "Performance Analysis of Ill-V Materials in a Double-Gate nano-MOSFET". *IEDM Tech. Dig.* pp. 113-116 (2007)

38. M. Zilli, P. Palestri, D. Esseni and L. Selmi, "On the experimental determination of channel back-scattering in nanoMOSFETs", *IEDM Tech. Dig.* p. 105. (2007)

39. P. Palestri, D. Esseni, S. Eminente, C. Fiegna, E. Sangiorgi and L.Selmi "Understanding Quasi-Ballistic Transport in Nano-MOSFETs: Part I – Scattering in the Channel and in the Drain," *IEEE Trans. Electron Devices*, **52**, no. 12, pp. 2727 – 2735, Dec. 2005.

40. S. Eminente, D. Esseni, P. Palestri, C. Fiegna, L. Selmi, E. Sangiorgi "Understanding Quasi-Ballistic Transport in Nano-MOSFETs: Part II – Technology Scaling Along the ITRS," *IEEE Trans. Electron Devices*, **52**, no. 12, pp. 2736–2743, Dec. 2005.

41. J. Saint Martin, A. Bournel, and P. Dollfus, "On the ballistic transport in nanometer-scaled DG MOSFETs", *IEEE Trans. Electron Devices*, **51**, no. 7, pp. 1148–1155, Jul. 2004.

42. P. Palestri, R. Clerc , D. Esseni, L. Lucci and L. Selmi, "Multi-Subband-Monte-Carlo investigation of the mean free path and of the kT layer in degenerated quasi ballistic nanoMOSFETs," in *IEDM Tech. Dig.* pp. 945–949 (2006)

43. M. V. Fischetti, T. P. O'Regan, S. Narayanan, C. Sachs, S. Jin, J. Kim, Y. Zhang, "Theoretical Study of Some Physical Aspects of Electronic Transport in nMOSFETs at the 10-nm Gate-Length," *IEEE Trans. Electron Devices*, **54**(9), pp. 2116 – 2136, Sept. 2007

44. M. V. Fischetti, S. Jin, T.-W. Tang, P. Asbeck, Y. Taur, S. E. Laux, M. Rodwell and N. Sano "Scaling MOSFETs to 10 nm: Coulomb effects, source starvation, and virtual source model" *J. Comp. Elec.*, **8**(2), p. 60-77 (2009).

45. S. Jin; T. W. Tang; M. V. Fischetti, "Anatomy of Carrier Backscattering in Silicon Nanowire Transistors" in *Proc. 13th International Workshop Computational Electronics*, 2009 pp.1 – 4

46. J. P. McKelvey and J. C. Balogh, "Flux methods for the analysis of transport problems in semiconductors in the presence of electric fields" *Phys. Rev.*, **137**(5A), pp. 1555–1561, Mar. 1965.

47. E. F. Pulver and J. P. McKelvey, "Flux methods for transport problems in solids with nonconstant electric fields," *Phys. Rev.*, **149**(2), pp. 617–623, Sep. 1966.
48. A. Lochtefeld and D. A. Antoniadis, "On experimental determination of carrier velocity in deeply scaled NMOS: How close to the thermal limit?" *IEEE Electron Device Lett.*, **22**(2), pp. 95–97, Feb. 2001
49. A. Lochtefeld; I. J. Djomehri; G. Samudra; D. A. Antoniadis; "New insights into carrier transport in n-MOSFETs" *IBM J. Res. Dev.*, **46**(2.3), pp. 347 – 357 (2002)
50. M.-J. Chen, H.-T. Huang, K.-C. Huang, P.-N. Chen, C.-S. Chang, and C. Diaz, "Temperature dependent channel backscattering coefficients in nanoscale MOSFETs," in *IEDM Tech. Dig*, pp. 39–42 (2002)
51. V. Barral, T. Poiroux, J. Saint-Martin, D. Munteanu, J. Autran, and S. Deleonibus, "Experimental investigation on the quasi-ballistic transport - Part 1: Determination of a New Backscattering Coefficient Extraction Methodology," *IEEE Trans. Electron Devices*, **56** (3), pp. 408–419, Mar. 2009.
52. V. Barral, T. Poiroux, D. Munteanu, J. Autran, and S. Deleonibus, "Experimental investigation on the quasi-ballistic transport - Part 2: Backscattering coefficient extraction and link with the mobility," *IEEE Trans. Electron Devices*, **56** (3), pp. 420–430, Mar. 2009.
53. C. Jeong; D.A. Antoniadis, M. S. Lundstrom, "On Backscattering and Mobility in Nanoscale Silicon MOSFETs" *IEEE Trans. Electron Devices*, **56**(11), pp. 2762 – 2769, Nov. 2009.
54. A. Cros, K. Romanjek, D. Fleury, S. Harrison, R. Cerutti, P. Coronel, B. Dumont, A. Pouydebasque, R. Wacquez, B. Duriez, R. Gwoziecki, F. Boeuf, H. Brut, G. Ghibaudo, T. Skotnicki, "Unexpected mobility degradation for very short devices: A new challenge for CMOS scaling", in *IEDM Tech. Dig.* pp. 399 (2006)
55. V. Barral, T. Poiroux, S. Barraud, F. Andrieu, O. Faynot, D. Munteanu, J. L. Autran, and S. Deleonibus, "Evidences on the Physical Origin of the Unexpected Transport Degradation in Ultimate n-FDSOI Devices *IEEE Trans. on Nanotechnology*, **8**(2) p. 167 (2009)
56. D. Fleury, G. Bidal, A. Cros, F. Boeuf, T. Skotnicki and G. Ghibaudo, "New Experimental Insight into Ballisticity of Transport in Strained Bulk MOSFETs", in *VLSI Tech. Dig.* 2009 pp. 16-17.
57. G. Ghibaudo, M. Mouis, L. Pham-Nguyen, K. Bennamane, I. Pappas, A. Cros, G. Bidal, D. Fleury, A. Claverie, G. Benassayag, P. F. Fazzini, C. Fenouillet-Beranger, S. Monfray, F. Boeuf, S. Cristoloveanu, T. Skotnicki, N. Collaert. "Electrical transport characterization of nano CMOS devices with ultra-thin silicon film", *Proc. of the International Workshop on Junction Technology*, pp. 58 – 63 (2009)
58. H. Wang, G. Gildenblat, "Scattering matrix based compact MOSFET model", in *IEDM Tech. Dig.*, pp. 125–128 (2002)
59. G. Gildenblat, "One-flux theory of a nonabsorbing barrier "*J. Appl. Phys.*, **91**(12), pp. 9883–9886 (2002).
60. S. Martinie, D. Munteanu, G. Le Carval, J.L. Autran, " Physics-Based Analytical Modeling of Quasi-Ballistic Transport in Double-Gate MOSFETs: From Device to Circuit Operation "*IEEE Trans. Electron Devices*, **56**(11), pp. 2692–2702 (2009)
61. A. Khakifirooz, D. Antoniadis, "MOSFET performance scaling—Part I: historical trends," *IEEE Trans. Electron Devices* **55**, p 1391-1400 (2008).
62. A. Khakifirooz, D. Antoniadis, "MOSFET performance scaling—Part II: future directions," *IEEE Trans. Electron Devices* **55**, p 1401-1408, (2008).
63. M. Lundstrom "Nanotransistors: A Bottom-Up View", *Proceeding of the ESSDERC*, pp. 33 - 40 (2006)
64. A. Rahman, M. Lundstrom, "A Compact Scattering Model for the Nanoscale Double-Gate MOSFET" *IEEE Trans. Electron Devices* **49**, p 481-489, (2002)

65. R. Clerc, P. Palestri, L. Selmi, "On the Physical Understanding of the kT-Layer Concept in Quasi Ballistic Regime of Transport in Nanoscale Devices," *IEEE Trans. Electron Devices*, **53**(7), pp. 1634 – 1639, (2006).

66. F. Assad, K. Banoo, M. Lundstrom, « The Drift Diffusion Equation revisited » *Solid State Electronics,* **42** p. 283 (1998).

67. J. H. Rhew, M. Lundstrom, « Drift Diffusion equation for ballistic transport in nanoscale metal-oxide-semiconductor field effect transistors » *J. Appl. Phys.*, **92** p. 5196-5202 (2002).

68. M. Lundstrom and J.-H. Rhew, "A Landauer Approach to Nanoscale MOSFETs", *J. Comp. Elec.* **1**: pp. 481–489, 2002.

69. R. Clerc, P. Palestri, L. Selmi, G. Ghibaudo "Back-Scattering in Quasi Ballistic NanoMOSFETs: The role of Non Thermal Carrier Distributions", *proc. of the ULIS conference*, p. 125 (2008). Submitted to *IEEE Trans. Electron Devices*.

70. E. Gnani, A. Gnudi, S. Reggiani, and G. Baccarani, Quasi-Ballistic Transport in Nanowire Field-Effect Transistors, *IEEE Trans. Electron Devices*, vol. 55, no 11, pp. 2918 – 2930, Nov. 2008.

71. D. Ponton, L. Lucci, P. Palestri, D. Esseni and L. Selmi, "Assessment of the Impact of Biaxial Strain on the Drain Current of Decanometric n-MOSFET", in *Proc. ESSDERC*, 2006, pp.166-169.

72. M. Lundstrom, "On the mobility versus drain current relation for a nanoscale MOSFET", *IEEE Electron Device Lett.*, **22**, pp. 293-295 (2001)

73. I. Pappas, G. Ghibaudo, C.A. Dimitriadis, C. Fenouillet-Béranger, « Backscattering coefficient and drift-diffusion mobility extraction in short channel MOS devices», *Solid State Electronics*, **53**, pp. 54–56 (2009).

74. M. Shur, "Low ballistic mobility in submicron HEMTs", *IEEE Electron Device Lett.*, **23**, pp. 511-513 (2002).

75. F. J. Wang and M. Lundstrom, "Ballistic transport in high electron mobility transistors", *IEEE Trans on Electron devices*, **50**, pp. 1604-1606 (2003).

76. G. Ghibaudo, "A new method for the extraction of MOSFET parameters", *Electronics Lett.*, **24**, pp. 543-544 (1988).

77. K. Romanjek, F. Andrieu, T. Ernst, and G. Ghibaudo, "Improved split C(V) method for effective mobility extraction in sub 0.1 µm Si MOSFETs", *IEEE Electron Device Lett.*, **25**, pp. 583-585 (2004).

78. F. Lime, G. Ghibaudo, F. Andrieu, J. Derix, F. Boeuf, T. Skotnicki, "Low temperature characterization of effective mobility in uniaxially and biaxially strained nMOSFETs", *Solid State Electronics*, **50**, pp. 644-648 (2003)

79. G. Bidal, D. Fleury, G. Ghibaudo, F. Boeuf and T. Skotnicki, "Guidelines for MOSFET Device Optimization accounting for L-dependent Mobility Degradation", in Proc. of Silicon Nanoelectronics Workshop, 2009.

PHYSICS BASED ANALYTICAL MODELING OF NANOSCALE MULTIGATE MOSFETs

TOR A. FJELDLY and UDIT MONGA

Department of electronics and Telecommunication
Norwegian University of Science and Technology, NO-7491 Trondheim, Norway,
and UNIK-University Graduate Center, Kjeller, Norway
torfj@unik.no

Various physics based modeling schemes for multigate MOSFETs are presented. In all cases, the models are derived from an analysis of the device body electrostatics in terms of two- or three-dimensional Laplace's and Poisson's equations, where short-channel and scaling effects are implicitly accounted for. Thus a comprehensive modeling framework is derived for the subthreshold electrostatics of double-gate MOSFETs based on a conformal mapping analysis of the potential distribution in the device body arising from the inter-electrode capacitive coupling. This technique is also applied to the circular gate-all-around MOSFET by utilizing the symmetry properties of this device. For both these devices, the modeling is extended to include the strong inversion regime by a self-consistent procedure that simultaneously allows the calculation of the quasi-Fermi potential distribution, the drain current, and the intrinsic capacitances. In an alternative modeling framework, covering a wide range of multigate devices in a unified manner, the potential distribution is derived from a select set of isomorphic trial functions that reflect the geometry and symmetry properties of the devices. Modeling parameters used are self-consistently determined by imposing boundary conditions associated with Laplace's or Poisson's equation. Finally, the effects of quantum mechanical confinement are discussed for ultra thin body devices. The results of the modeling are in excellent agreement with numerical simulations.

Keywords: Multigate MOSFETs; device modeling; nanoscale devices.

1. Introduction

Multigate MOSFETs (MugFETs) in the nanoscale range have been identified as strong candidates for replacing the conventional bulk MOSFET in the coming years, to meet the ever growing demand for high-speed, low-power CMOS circuitry.[1] A major impetus for this advance is the improved gate control and the concomitant reduction in short-channel behavior offered by these device designs. As an integral part of this development, precise compact models of the devices are needed for implementation in circuit simulators and circuit design tools. To achieve the required precision, the higher dimensionality of the potential and inversion charge distributions have to be taken into account. Much of the literature in this area has been concentrated on undoped or long-channel devices.[2,3,4,5,6,7,8,9,10,11] However, as device channel lengths are steadily reduced to increase speed, there is always a need for precise descriptions of the short-channel effects (SCEs) associated with these devices.

In nanoscale MugFETs, the electron barrier topology is a critical factor when determining conduction paths and currents in the devices. So-called volume inversion is a well-known phenomenon in MugFETs, where the inversion charge in the subthreshold regime tends to collect in the body interior and to shift towards the gates above threshold. A prerequisite for describing this complicated behavior is to include two- or three dimensional (2D or 3D) effects in the models, preferably based on self-consistent solutions of the electrostatics in the device. In such an approach, short-channel effects and scaling properties will be intrinsic to the model, which, accordingly, will require only a minimal parameter set of clear physical origin.

For the double-gate (DG) MOSFET, the basic modeling problem is to obtain an analytical or semi-analytical solution of a 2D Poisson's equation where the contacts (source, drain and the two gates) and the dielectric gaps at the corners of the body cross-section define the boundary. According to the superposition principle, Poisson's equation can be separated into a 2D Laplace equation for the inter-electrode capacitive coupling, and a remainder which involves the potential distribution established by the body charges. The Laplace problem for the DG MOSFET can be solved in different ways. One possibility is to perform a full Fourier expansion of the potential or by using a low-order truncation.[12,13,14] An alternative approach is to apply a conformal mapping technique[15], which was first used for classical, long-channel MOSFETs.[16,17] Later, this technique was enhanced and applied to the subthreshold regime of short-channel, nanoscale DG MOSFETs[18,19,20,21]. In Section 2, we present modeling results on electrostatics, drain current, and intrinsic capacitances based on this technique applied to a wider range of operation, including the strong inversion regime.[22,23,24,25,26,27,28]

The gate-all-around (GAA) circular, cylindrical gate (CirG) MOSFET, which is basically a 3D structure, cannot be analyzed directly using the conformal mapping technique. One possibility is to solve Laplace's equation in cylindrical coordinates by means of a series expansion in Bessel functions.[29] As an alternative, we have applied the analytical results from the DG structure, together with the inherent symmetry properties of the CirG device, to obtain an accurate description of the device both in subthreshold and in strong inversion.[24,26,27,30,31],

The above modeling techniques based on conformal mapping analysis provides a high degree of precision. Even though this analysis leads to important analytical results for the electrostatics in the subthreshold regime, the technique requires a Scwartz-Christoffel transformation[15] involving complex elliptic integral functions, which represents an added computational overhead when used in circuit simulations. However, based on results from this technique, we have developed an alternative, analytical modeling framework covering a wide range of multigate devices.[32,33,34] In this framework, both subthreshold and strong inversion electrostatics is derived from a set of isomorphic trial functions that reflect the geometric and symmetry properties of the devices considered. The modeling parameters used are self-consistently determined applying Laplace's or Poisson's equations and appropriate boundary conditions. This modeling technique, which represents an excellent basis for a unified modeling framework for multigate MOSFETs, is presented in Section 4.

Finally, in Section 5, we discuss some important issues related to the effects of quantum mechanical confinement in ultra-thin body DG and CirG MOSFETs.

For the devices considered here, the permittivities used are $\varepsilon_{si} = 11.8\ \varepsilon_0$ for silicon and $\varepsilon_{ox} = 7\varepsilon_0$ for the gate insulator, where ε_0 is the vacuum permittivity. The doping density of the *p*-type silicon body is $N_a = 1\text{x}10^{15}\text{cm}^{-3}$. As gate material, we use a near-midgap metal with the work function 4.53 eV. Idealized Schottky contacts with a work function of 4.17 eV (corresponding to that of n^+ silicon) are assumed for the source and drain. This ensures equipotential surfaces on all the device contacts. The device dimensions considered are such that a classical treatment of the electron distribution is justified,[35,36] unless specified otherwise. The same assumption is also made in the numerical simulations using the ATLAS semiconductor simulator from Silvaco Int.

2. Modeling of DG MOSFETs Based on Conformal Mapping Techniques

In an electronic device such as a multigate MOSFET, a prerequisite for a reliable modeling of its intrinsic properties is to accurately describe the electrostatics of the device body for the required range of biasing conditions. Such a description must be based on Poisson's equation, which relates the local electrostatic potential φ to the space charge density ρ as follows

$$\frac{\partial^2 \varphi}{\partial x^2} + \frac{\partial^2 \varphi}{\partial y^2} + \frac{\partial^2 \varphi}{\partial z^2} = -\frac{\rho(x, y, z)}{\varepsilon}, \tag{1}$$

where ε is the dielectric constant. The solution of (1) depends on the boundary conditions of the device body, such as the properties and biasing conditions of the electrodes (Dirichlet boundary conditions). An advantage of nanoscale MugFETs, however, is that low- or undoped silicon bodies are preferred. Therefore, owing to the small volume involved, the contribution to the potential by doping concentrations of up to about 10^{16} cm^{-3} is negligible compared to the combined contributions of the inter-electrode capacitive coupling (see below) and the charge carriers. Nonetheless, an important function of the doping is to allow adjustment of the body Fermi potential.

Generally, the solution of (1) can be separated into two additive parts (superposition principle), one of which is the particular solution φ^{LP} of the Laplace equation,

$$\frac{\partial^2 \varphi^{LP}}{\partial x^2} + \frac{\partial^2 \varphi^{LP}}{\partial y^2} + \frac{\partial^2 \varphi^{LP}}{\partial z^2} = 0. \tag{2}$$

This solution describes the capacitive coupling between the electrodes. In subthreshold conditions where the minority carrier charge is small or negligible, this inter-electrode potential will be dominant. The contribution φ^Q associated with the minority carrier charge density n can then be calculated self-consistently from the residual part of (1), i.e.,

$$\frac{\partial^2 \varphi^Q}{\partial x^2} + \frac{\partial^2 \varphi^Q}{\partial y^2} + \frac{\partial^2 \varphi^Q}{\partial z^2} = -\frac{qn(x, y, z)}{\varepsilon}, \tag{3}$$

resulting in the total potential $\varphi = \varphi^Q + \varphi^{LP}$.

For the DG MOSFET, we can assume that the device width is much larger than the thickness, allowing us to consider a 2D cross section of the device as shown in Fig. 1, in which case also (1) to (3) become two-dimensional. The geometric dimensions of the DG device considered are $L = 25$ nm, $t_{si} = 12$ nm, and $t_{ox} = 1.6$ nm. Other device parameter values are indicated in Section 1.

2.1. *Conformal Mapping*

The use of conformal mapping is a powerful technique for solving 2D electrostatic problems, such as the inter-electrode electrostatics of the rectangular cross-section of the DG device shown in Fig. 1a, including the silicon body and the gate oxides. As indicated in Section 1, this problem can be described based on the 2D Laplace equation, using the potentials at the gate, source, drain, and the insulator gaps at the corners as boundary conditions[18-21]. However, to simplify the modeling, we replace the gate insulator by an electrostatically equivalent silicon layer of thickness $t'_{ox} = t_{ox}\varepsilon_{si}/\varepsilon_{ox}$, where t_{ox} is the true gate insulator thickness. We then define an extended silicon body of length L, relative permeability ε_{si}, and thickness $t_{si} + 2t'_{ox}$, where t_{si} is the thickness of the true silicon body. We have found that this procedure is quite precise for all practical device geometries and gate insulator materials used in DG devices. The boundary of the extended body is then defined as its interfaces with the four electrodes and the insulator gaps at the four corners.

The inter-electrode potential distribution in the extended body is derived by performing a conformal mapping of this body from the normal (x,y) plane to the upper half-plane of a complex (u,iv) plane as indicated in Fig. 1. This mapping is defined by the following Schwartz-Christoffel transformation,[15,19]

$$z = x + iy = \frac{L}{2} \frac{F(k,w)}{K(k)} \tag{4}$$

where $w = u + iv$,

$$F(k,w) = \int_0^w \frac{dw'}{\sqrt{\left(1 - w'^2\right)\left(1 - k^2 w'^2\right)}} \tag{5}$$

is the complex elliptic integral of the first kind, $K(k) = F(k,1)$ is the corresponding complete elliptic integral, and the modulus k is a constant between 0 and 1 determined by the extended body aspect ratio from,[19]

$$\frac{K\left(\sqrt{1-k^2}\right)}{2K(k)} = \frac{t_{si} + 2t'_{ox}}{L} . \tag{6}$$

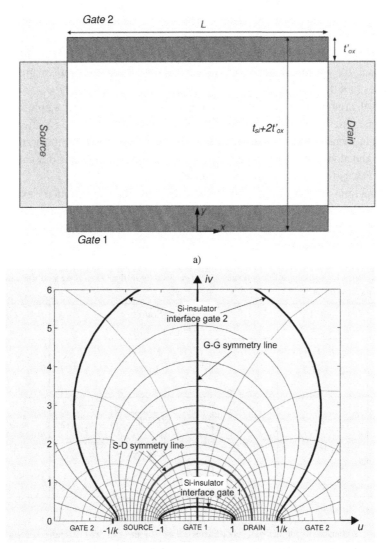

Fig. 1. Schematic view of the DG MOSFET cross section in a) the (x,y) plane and in b) the (u,iv) plane. In (b) the u axis represents the device boundary, the iv axis represents the gate-to-gate symmetry axis, the semi-circle represents the source-to-drain symmetry axis, and the inner and outer bold curves indicate the silicon-insulator interfaces at Gate 1 and Gate 2, respectively. The thin curves in (b) indicate how a rectangular grid in the (x,y) plane maps into the (u,iv) plane.

We note that in the (u,iv) plane shown in Fig. 1b, $u = 0$, $v = 0$ corresponds to the middle point on the lower gate contact (Gate1) in Fig. 1a, which is defined as the origin ($x = 0$, $y = 0$) in the (x,y) plane. Moreover, the boundary of the extended body is mapped into the real u axis, the four corners map into $u = \pm 1$ and $u = \pm 1/k$. Gate 1 extends from $u = -1$ to $u = +1$ and Gate 2 extends from $u = 1/k$ to ∞ and from $u = -1/k$ to $-\infty$. The limits at $u =$

$\pm\infty$ correspond to the center of Gate 2. The iv-axis represents the gate-to-gate (G-G) symmetry axis through the body center where $v = \infty$ also corresponds to the center of Gate 2. The source-to-drain (S-D) symmetry axis is represented by a half-circle of radius $1/\sqrt{k}$ about the origin in the (u,iv) plane and the body center is mapped into the coordinates $(u = 0, v = 1/\sqrt{k})$.

For real arguments in the standard range $0 \le u \le 1$, $v = 0$, ample approximate expressions, series expansions, and iteration routines exist for $F(k,u)$.[37] $F(k,w)$ can also be expressed in terms of the standard elliptic integral for other values of the argument on the boundary and along the two symmetry lines. In addition, routines exist for calculating the values of $F(k,w)$ for general complex arguments w.

The functions for the mapping of the boundary and the two symmetry axes are given by the expressions in Table 1 and 2, respectively, in terms of the standard elliptic integral obtained from expressions (4) and (5) (note that $F(k,-u) = -F(k,u)$).[20] These mapping functions are illustrated in Fig. 2.

Table 1. Functions for mapping of coordinates on the real u-axis in the (u,iv) plane into the extended body boundary in the (x,y) plane.

	Gate1 $(-1 \le u \le 1)$	Drain, Source $(1 < |u| \le 1/k)$,	Gate2 $(1/k < |u| \le \infty)$
x	$\dfrac{L}{2}\dfrac{F(k,u)}{K(k)}$	$\pm\dfrac{L}{2}$	$\dfrac{L}{2}\dfrac{F\left(k,\dfrac{1}{ku}\right)}{K(k)}$
y	0	$\left(t_{Si} + 2t'_{ox}\right)\left[1 - \dfrac{F\left(\sqrt{1-k^2},\sqrt{\dfrac{1-k^2u^2}{1-k^2}}\right)}{K\left(\sqrt{1-k^2}\right)}\right]$	$t_{si} + 2t'_{ox}$

Table 2. Functions for mapping coordinates related to the gate-to-gate and source-to-drain symmetry axes in the (u,iv) plane into those of the extended device body in the (x,y) plane.

	Gate-to-gate symmetry axis $(u = 0, 0 \le v \le \infty)$	Source-to-drain symmetry axis $(|u| \le 1/\sqrt{k}, v = \sqrt{1/k - u^2})$
x	0	$\dfrac{L}{2}\dfrac{F\left(\dfrac{2\sqrt{k}}{1+k},\sqrt{k}u\right)}{K\left(\dfrac{2\sqrt{k}}{1+k}\right)}$
y	$\left(t_{Si} + 2t'_{ox}\right)\dfrac{F\left(\sqrt{1-k^2},\dfrac{v}{\sqrt{1+v^2}}\right)}{K\left(\sqrt{1-k^2}\right)}$	$\dfrac{t_{si}}{2} + t'_{ox}$

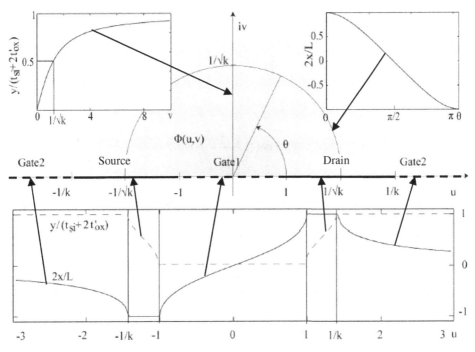

Fig. 2. The extended body of the DG MOSFET mapped into the upper half of the (u,iv)-plane. The insets show the mapping functions for the u-axis (lower), the iv-axis (upper left) and the semicircle of radius $1/\sqrt{k}$ (upper right). These represent the boundary, the gate-to-gate symmetry axis, and the source-to-drain symmetry axis, respectively. (After Ref. 20)

Note that the inverse of the transform (4), i.e., from the (x,y) plane to the (u,iv) plane, can be expressed in terms of Jacobi elliptic functions.[23,36]

2.2. *Inter-Electrode and Subthreshold Electrostatics in DG MOSFETs*

The inter-electrode electrostatics of the DG device based on the 2D Laplace equation can be described analytically in the (u, iv) plane, using the boundary conditions indicated above.[18-21] This solution, which is dominant in the subthreshold regime, can be obtained from[15]

$$\varphi_{DG}^{LP}(u,v) = \frac{v}{\pi} \int_{-\infty}^{\infty} \frac{\varphi_{DG}(u',0)}{(u-u')^2 + v^2} du', \qquad (7)$$

where $\varphi_{DG}(u',0)$ is the electrostatic potential along the entire boundary. The major contributions to this integral come from the four equipotential contacts and minor terms come from the insulator gaps at the corners. In the limit of zero insulator thickness, (7) results in the following analytical expression for the potential distribution[19]

$$\varphi_{DG}^{LP}(u,v)=\frac{1}{\pi}\left\{ \begin{array}{l} \left(V_{gs2}-V_{FB}\right)\left[\pi-\tan^{-1}\left(\dfrac{1-ku}{kv}\right)-\tan^{-1}\left(\dfrac{1+ku}{kv}\right)\right] \\[2mm] +\left(V_{gs1}-V_{FB}\right)\left[\tan^{-1}\left(\dfrac{1-u}{v}\right)+\tan^{-1}\left(\dfrac{1+u}{v}\right)\right] \\[2mm] +V_{bi}\left[\tan^{-1}\left(\dfrac{1-ku}{kv}\right)-\tan^{-1}\left(\dfrac{1-u}{v}\right)\right] \\[2mm] +\left(V_{bi}+V_{ds}\right)\left[\tan^{-1}\left(\dfrac{1+ku}{kv}\right)-\tan^{-1}\left(\dfrac{1+u}{v}\right)\right] \end{array}\right\}, \tag{8}$$

where V_{gs1} and V_{gs2} are the potentials of Gate 1 and Gate 2, respectively, referred to the source contact, V_{FB} is the flat-band voltage between the gates and the silicon body, V_{bi} is the built-in voltage of the source and drain, and V_{ds} is the drain-source voltage.

Along the G-G and S-D symmetry axes, respectively, (8) simplifies as follows for symmetric gate biasing ($V_{gs} = V_{gs1} = V_{gs2}$),

$$\varphi_{DG}^{LP}(0,v)=V_{gs}-V_{FB}-\frac{1}{\pi}\left(2V_{gs}-2V_{FB}-2V_{bi}-V_{ds}\right)\left[\tan^{-1}\left(\frac{1}{kv}\right)-\tan^{-1}\left(\frac{1}{v}\right)\right], \tag{9}$$

$$\varphi_{DG}^{LP}\left(u,\frac{1}{\sqrt{1/k-u^2}}\right)=V_{gs}-V_{FB}-\frac{1}{\pi}\left\{\left(V_{gs}-V_{FB}-V_{bi}\right)\left[\tan^{-1}\left(\frac{1/k-u}{\sqrt{1/k-u^2}}\right)-\tan^{-1}\left(\frac{1-u}{\sqrt{1/k-u^2}}\right)\right]\right. \tag{10}$$
$$\left. +\left(V_{gs}-V_{FB}-V_{bi}-V_{ds}\right)\left[\tan^{-1}\left(\frac{1/k+u}{\sqrt{1/k-u^2}}\right)-\tan^{-1}\left(\frac{1+u}{\sqrt{1/k-u^2}}\right)\right]\right\}$$

and the potential at the device center becomes,

$$\varphi_{DG}^{LP}\left(0,\frac{1}{\sqrt{k}}\right)=V_{gs}-V_{FB}-\left(V_{gs}-V_{FB}-V_{bi}-\frac{V_{ds}}{2}\right)\left[1-\frac{4}{\pi}\tan^{-1}\left(\sqrt{k}\right)\right]. \tag{11}$$

Moreover, taking the derivative with respect u of φ_{DG}^{LP} in (10) and setting it equal to zero, we obtain the following expression for the location of the barrier maximum along the S-D symmetry axis,[23,38]

$$u_m=-\frac{(1+k)V_{ds}}{2k\left(2V_{gs}-2V_{FB}-2V_{bi}-V_{ds}\right)}, \qquad v_m=-\frac{1}{\sqrt{1/k-u_m^2}}. \tag{12}$$

The corresponding potential is obtained by using $u = u_m$ in (10).

Using the mapping function of (4) (or its inverse transform), the potential profile can be mapped back to the (x,y) plane. Figure 3 shows the 2D inter-electrode potential distribution of the DG MOSFET in the (x,y) plane modeled using (8) with $V_{ds} = 0.2$ V and $V_{gs1} = V_{gs2} = 0$ V, which correspond to subthreshold conditions.[23] For $V_{ds} = 0$ V, $\varphi_{DG}^{LP}(x,y)$ has a saddle point at the device center, corresponding to the minimum barrier energy

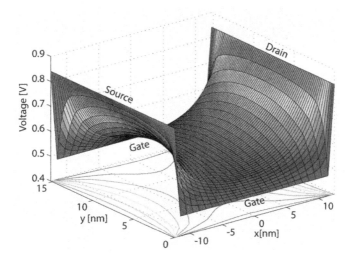

Fig. 3. Symmetric DG MOSFET subthreshold inter-electrode potential distribution for $V_{ds} = V_{gs} = 0$ V transformed to the (x,y)-plane. The results are obtained from (8) assuming a thin insulator. Device dimensions: $L = 25$ nm, $t_{si} = 12$ nm, and $t_{ox} = 1.6$ nm.

for electron conduction between source and drain. With increasing V_{ds}, the barrier minimum is steadily lowered and shifts towards source. This drain-induced barrier lowering (DIBL) as well as other short-channel effects is inherent to the present formalism. Increasing the symmetric gate biasing, the gate-to-gate barrier energy profile is lowered and flattens, and the barrier minimum eventually shifts to the silicon-insulator interfaces at the transition voltage V_T. At this stage, the induced electron density will strongly influence the device electrostatics, requiring a self-consistent analysis.

As indicated earlier, the above formalism is also applicable for asymmetric gate biasing or even asymmetric gate structures. Figure 4 shows an example of the potential distribution in the latter, where the mid-gap gate metal has been replaced by p$^+$ polysilicon in Gate 2.

2.2.1. *Corner correction*

The results above were derived for the limit of zero gate insulator thickness. This approximation works quite well for evaluating the potential distribution in the body interior. Typical errors compared to numerical device simulations are on the order 10 mV for the device considered. However, for the charge density near the barrier maximum and for the subthreshold current, this easily translates into an error of a factor of about 1.5 to 2. Close to the corners, the errors in charge density may be quite a lot higher, which mostly affects the precision in the modeling of capacitances.

A relatively simple way of correcting the interior body potential and the drain current is to approximate the potential distribution through the insulator at each corner by a step function, defined by extending both the gate interface potential $V_{gs} - V_{FB}$ and the source/

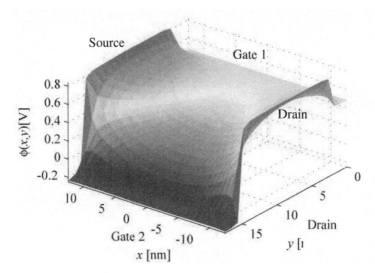

Fig. 4. Subthreshold inter-electrode potential distribution for asymmetric DG MOSFET with p⁺ polysilicon as gate material for Gate 2. Bias voltages: $V_{ds} = V_{gs} = 0$ V. The results were obtained from (8) assuming a thin insulator and are mapped in to the (x,y)-plane using the inverse transformation of (4). (After Ref. 23.)

drain potentials along the boundary to a position of about $0.88t'_{ox}$ from the gate.[23,24] However, this procedure tends to worsen the description close to the corners.

A more precise approximation for the corner regions is obtained by analyzing the potential of the boundary sections across the four insulator gaps, again using conformal mapping. Since the distance between the corners is quite large compared to the insulator thickness, such an analysis can be simplified by considering only one corner and assuming that the adjoining gate and source/drain regions have an infinitely extent. This structure is indicated in Fig. 5, where the new one-corner coordinates $z_{1c} = y_{1c} + ix_{1c}$ for the boundary are indicated. Here, we also assume that the overlap between gate-source-drain (lower left in Fig. 5) is larger than the insulator thickness, in which case we take this overlap region to be infinitely long for the purpose of making the analysis manageable. As before, we replace the oxide layer by an electrostatically equivalent silicon layer. This approach is not quite precise very close to the corner, but the overall result is satisfactory.

Considering the region in Fig. 5 consisting of the quadrant to the right and the oxide layer extending to the left, all enclosed by the by the gate and source (or drain) electrodes (bold black boundary lines), this gives rise to the following Schwartz-Christoffel transformation:[27,39]

$$z_{1c} = y_{1c} + ix_{1c} = \frac{2t'_{ox}}{\pi}\left[\sqrt{w_{1c}-1} - \tan^{-1}\left(\sqrt{w_{1c}-1}\right)\right] + t'_{ox}. \tag{13}$$

Here $w_{1c} = u_{1c} + iv_{1c}$ are the coordinates of the complex (u_{1c}, iv_{1c}) half-plane into which the one-corner geometry is mapped. Applying expression (4) in this half-plane, we obtain the

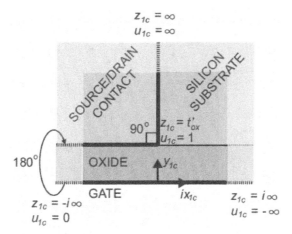

Fig. 5. One-corner structure extending to infinity in the upper half-plane. The boundary of the area analyzed is indated by fat lines. Important boundary coordinates z_{1c} are indicated along with their corresponding transformed u_{1c} values. (After Ref. 27.)

following potential distribution,[27]

$$\varphi_{1c}\left(u_{1c}, v_{1c}\right) = \frac{1}{2}\left(V_G + V_{S/D}\right) - \frac{1}{\pi}\left(V_G - V_{S/D}\right)\tan^{-1}\left(\frac{u_{1c}}{v_{1c}}\right), \tag{14}$$

where V_G, V_S and V_D are the gate, source and drain voltages, respectively. We note that the electrostatics in this case is described by radial equipotential lines emanating from the origin and semi-circular electrical field lines, as shown in Fig. 6a. The corresponding distributions in the (x_{1c}, y_{1c})-plane are shown in Fig. 6b.

The dashed bold curve in Figs. 6a and 6b represents the insulator gaps in the boundary of the DG extended device body (see Fig. 1) reaching from the source/drain corner at $(x_{1c} = 0, y_{1c} = t'_{ox})$ or $(u_{1c} = 1, v_{1c} = 0)$ to the gate at $(x_{1c} = 0, y_{1c} = 0)$ or $(u_{1c} = u_o = -0.439, v_{1c} = 0)$. The continuous bold curve shows the field line that emanates from the source/drain corner and terminates on the gate at $(x_{1c} = 0.339t'_{ox}, y_{1c} = 0)$ or $(u_{1c} = -1, v_{1c} = 0)$. Note that the origin in Fig 6b corresponds to $(x_{1c} = -\infty, y = 0)$ in Fig. 6b.

The inset in Fig. 6b indicates the shape of the potential profile over the insulator gap obtained from this analysis. The following expression closely approximates this profile,[27]

$$\varphi_{ox} = Ay_{1c}^6 + By_{1c} + C, \tag{15}$$

where the parameters A, B and C are determined by the gate and source/drain potentials and the derivative of φ_{ox} at the gate,

$$\frac{d\varphi_{ox}}{dy_{1c}} = \frac{V_{S/D} - V_G}{t'_{ox}\sqrt{1 + |u_o|}}, \tag{16}$$

where u_o is indicated in Fig. 6a.

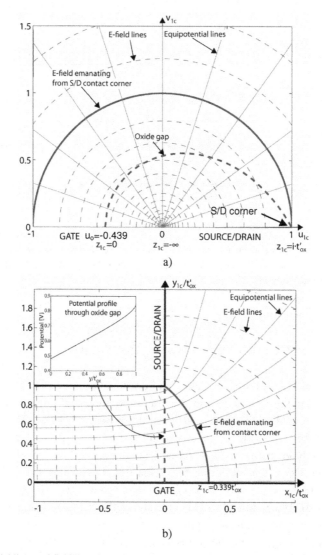

Fig. 6. Equipotential lines and field lines from the one-corner analysis in the (u_{1c}, iv_{1c})-plane (a), and the same in the (x_{1c}, y_{1c})-plane (b). The bold solid curve indicates the field line emanating from the corner. The bold dashed curve shows the insulator gap part of the DG boundary. The inset in b) indicates the potential profile along the latter. (After Refs. 27 and 38.)

When applying φ_{ox} of (15) to the four insulator gaps of the DG extended body, it has to be transformed into the (u, iv)-plane of the full DG device and applied to the appropriate boundary sections of the integral in (7). This approximate procedure, which is detailed in Ref. 27, results in the updated potential distribution shown in Fig. 7, which should be compared with that of Fig. 3. Note especially the adjusted shape of the distribution in the corner regions.

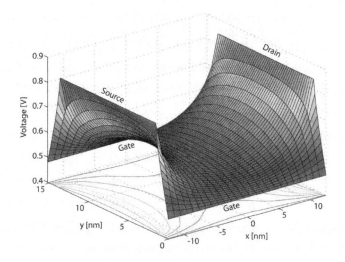

Fig. 7. DG MOSFET subthreshold inter-electrode potential distribution including the corner correction. Same device and biasing as in Fig. 3. (After Ref. 27.)

2.2.2. *Effect of subthreshold minority carriers near source and drain*

For a precise modeling of the subthreshold potential, we also have to include the effects of the small electron population near source and drain. A low-order correction can be made self-consistently by applying a 1D Poisson's equation to these regions, provided that the electric fields E_S and E_D at the source and drain are dominated by the fields associated with the inter-electrode electrostatics i.e., E_S^{LP} and E_n^{LP}. For the center of source and drain, we obtain the following analytical expressions from (4), (5), and (10),[23,24]

$$
\begin{aligned}
E_{S/D}^{LP} &= -\frac{d\varphi_{S/D}^{LP}}{dx}\bigg|_{x=\pm L/2} = -\frac{d\varphi_{S/D}^{LP}}{dx}\frac{dx}{du}\bigg|_{x=\pm L/2} \\
&= \pm\frac{8}{\pi L}\frac{\sqrt{k}}{1+k}K\!\left(\frac{2\sqrt{k}}{1+k}\right)\!\left[\left(V_{bi}-V_{gs}-V_{FB}\right)\mp\frac{\left(1\mp\sqrt{k}\right)^2}{4\sqrt{k}}V_{ds}\right].
\end{aligned}
$$

(17)

Here the upper and lower signs refer to the source and drain, respectively. This result is a fairly good approximation for a wide central portion of the contact surfaces.

Using Boltzmann statistics for the electrons, we obtain the following approximate expression for the total electric field at source and drain center,

$$
E_{S/D} \approx E_{S/D}^{LP} \pm \frac{qN_C}{\varepsilon_{si}}\int_0^\infty \exp\!\left(-\frac{\left|E_{S/D}^{LP}\right|x'}{V_{th}}\right)dx' \approx E_{S/D}^{LP} + \frac{qN_C V_{th}}{\varepsilon_{si}E_{S/D}^{LP}}
$$

(18)

Here, V_{th} is the thermal voltage and $N_C \approx \left(n_i^2/N_a\right)\exp\left(V_{bi}/V_{th}\right)$ is the electron concentration at the contact-body interface. We observe from (17) that the magnitude of E_D^{LP} increases with increasing drain bias while E_S^{LP} is relatively unaffected, except at relatively short gate lengths (see Section 2.2.3). Moreover, according to (18), the effect of the inversion electrons decreases with increasing drain bias near the drain. When the device is driven deeper into subthreshold, we observe an increase in the magnitude of both E_S^{LP} and E_D^{LP}, and a decrease of the free carrier effect on both sides.

The effect of the inversion charge may also be interpreted as an adjustment of the potential boundary conditions at source and drain by

$$\Delta\varphi_{S/D} \approx V_{th}\left(\frac{E_{S/D}}{E_{S/D}^{LP}}-1\right), \tag{19}$$

which corresponds to a fraction of a thermal voltage. Although small, $\Delta\varphi_{S/D}$ may have a non-negligible effect on the charge density in the body interior.

2.2.3. *Verification of subthreshold electrostatics*

The above model for the DG MOSFET subthreshold electrostatics, including the corrections terms, are compared with numerical simulations in Figs. 8 and 9, using the ATLAS circuit simulator. The device specifications are the same as that of Fig. 3.

Figure 8 shows the potential distribution along the S-D and G-G symmetry lines for $V_{ds} = V_{gs} = 0$ V. We note the good match between the curves where they cross at the device center. In Fig. 9 are shown potential profiles along the S-D symmetry axis for $V_{gs} = 0$ V and different values of V_{ds}. The DIBL manifests itself as an increase of the minimum potential and a shift towards the source of the location of the minimum with increasing V_{ds}.

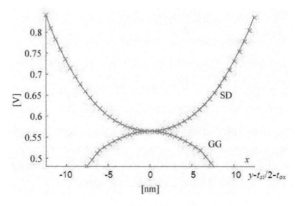

Fig. 8. Subthreshold potential distribution along S-D symmetry axis (upper curve, x-direction) and along the G-G symmetry axis (lower curve, y-direction) for $V_{ds} = V_{gs} = 0$ V. Solid lines indicate modeling results and symbols show numerical simulations. In the G-G curve is shown the potential in the true insulator, indicated by kinks in the curve at the silicon-oxide interface ($\pm$ 6 nm). (After Ref. 23.)

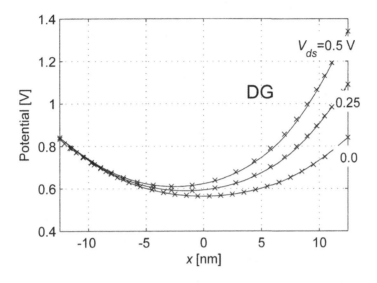

Fig. 9. Modeled (solid curves) and simulated (symbols) subthreshold potential profiles along the S-D symmetry axis of the DG MOSFET for $V_{gs} = 0$ V and $V_{ds} = 0$ V, 0.25 V, and 0.5 V. Drain-induced barrier lowering is indicated by the shift in location and value of the potential minimum with V_{ds}. (After Refs. 23 and 30.)

   In order to check the scaling properties of the present DG MOSFET model, the shift of the the potential minima away from the device center with increasing V_{ds} was calculated from (12) for devices with gate lengths ranging between 12.5 nm and 50 nm, keeping all other dimensions fixed at the values given above. As shown in Fig. 10, the modeled values agree very well with numerical simulation down to a gate length of at least 15 nm, demonstrating good scaling properties of the DG device model. For $L = 12.5$ nm, there is a significant deviation for $V_{ds} > 0.2$ V. This is likely due to a breakdown of the assumptions made in Section 2.2.2 on the relative magnitude of the influence of the inversion charge on the electrostatics near the source.

2.2.4. Subthreshold drain current

Based on the above analysis of the subthreshold electrostatics (neglecting the minority carrier charge), it is possible to derive an expression for the subthreshold drain current of DG MOSFETs. Using the drift-diffusion transport mechanism and assuming a constant mobility (no velocity saturation), this current can be written as

$$I_{dsub} = q\mu_n W \iint n(x, y)) \frac{dV_F}{dx} dy dx, \qquad (20)$$

Here the double integral runs over the length and height of the silicon body, W is the device width, μ_n is the electron mobility, $n(x,y)$ is the electron distribution, and $V_F(x)$ is the quasi-Fermi potential, which is assumed to be constant over any given cross-section perpendicular to the x-axis. Invoking current continuity along the channel and separating

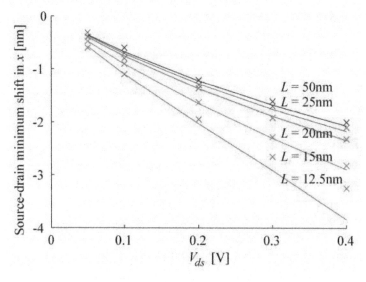

Fig. 10. Modeled (curves) and simulated (symbols) shift of potential minimum along the S-D axis with increasing V_{ds} for different gate lengths and constant silicon and insulator thicknesses. Gate bias: $V_{gs} = 0$ V. (After Ref. 23.)

(20) into a coordinate dependent and a V_F dependent part, we obtain the following version of the Pao-Sah expression[40]

$$I_{dsub} = \mu_n q n_s(x) W \frac{dV_F}{dx} = q\mu_n W V_{th} \frac{1 - \exp(-V_{ds}/V_{th})}{\int_{-L/2}^{L/2} \frac{dx}{n_{so}(x)}}, \qquad (21)$$

where $n_s(x)$ is the sheet density of electrons along the channel and

$$n_{so}(x) = \frac{n_i^2}{N_a} \int_{t_{ox}}^{t_{si}+t'_{ox}} \exp\left(\frac{\varphi^{LP}(x,y)}{V_{th}}\right) dy \qquad (22)$$

is the sheet density when taking the quasi-Fermi level in the body to be fixed to that of the source.

Figure 11 shows an example of subthreshold I_{dsub}-V_{ds} characteristics for the same device as discussed above. The modeling was done by solving the integrals in (21) and (22) numerically by means of a Simpsons formula, and the results are compared with numerical simulations (Atlas). The noticeable output conductance in saturation is an illustration of loss of gate control brought on by short-channel effects in the device considered.

We observe that the integrals in (21) and (22) don't have simple analytical solutions because of the functional form of $\varphi^{LP}(x,y)$. However, it has been shown that the inverse charge density $1/n_s(x)$ can be represented quite well by a normal distribution centered at the top of the barrier where $n_s(x)$ has its minimum value n_{sm},[41] i.e.,

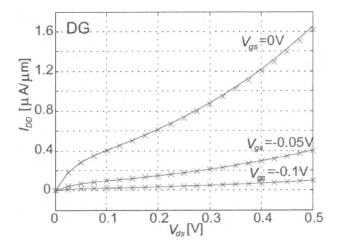

Fig. 11. Comparison of modeled (symbols) and simulated (solid curves) I_{dsub}-V_{ds} subthreshold characteristics for DG MOSFET. Same device as in Fig. 3. (After Ref. 27.)

$$\frac{1}{n_s(x)} = \frac{1}{n_{sm}} \exp\left[-\left(\frac{x-x_m}{\sigma}\right)^2\right]. \tag{23}$$

Here $x = x_{sm}$ is the position of the barrier maximum and $x_m \pm \sigma$ defines the positions where the charge has increased by a ratio of e compared to that of the top of the barrier.[23,40] Note that σ and n_{sm} are directly related to the equal and opposite curvature of the potential distribution in the x and y directions at the top of the barrier. Any error in the distribution beyond a few times σ away from x_m can be neglected for all practical cases.

Using the above results together with Boltzmann statistics for the electron distribution, n_{sm} can be expressed in terms of the error function as[42]

$$n_{sm} = \frac{n_i^2}{N_a}\left(\frac{t_{si}}{2} + t'_{ox}\right)\sqrt{\frac{\pi V_{th}}{\hat{\varphi}^{LP}(x_m)}} \exp\left[\frac{\varphi^{LP}(x_m)}{V_{th}}\right] Erf\left[\frac{t_{si}}{t_{si} + 2t'_{ox}}\sqrt{\frac{\hat{\varphi}^{LP}(x_m)}{V_{th}}}\right], \tag{24}$$

where $\hat{\varphi}^{LP}(x_m) = \varphi^{LP}(x_m) - V_{gs} + V_{FB}$ is the potential at $x = x_m$ on the D-S symmetry axis relative to the that at the gate boundary. The position of the barrier maximum is obtained from the conformal mapping analysis in terms of u_m in (12), from which x_m is obtained using the mapping function (4). For increased accuracy, the corrections discussed in Sections 2.2.2 should also be implemented.

Combining (21) to (24), we obtain the following expression for the drain current[40]:

$$I_{dsub} = \frac{2q\mu_n W V_{th} n_{sm}}{\sqrt{\pi}\sigma}\frac{1 - \exp(-V_{ds}/V_{th})}{Erf\left(\frac{L/2 + x_m}{\sigma}\right) + Erf\left(\frac{L/2 - x_m}{\sigma}\right)}. \tag{25}$$

The modeled current compared with numerical simulations for DG MOSFETS of lengths $L = 25$ nm and 50 nm are shown in Fig. 12. The other parameters are the same as for the DG MOSFET considered in Section 2.2).

An important quality parameter for MOSFETs is the subthreshold slope, which can be determined as follows[40]

$$S = \frac{\ln(10)I_{dsub}}{dI_{dsub}/dV_{gs}}. \tag{26}$$

Hence, an analytical expression for S may be obtained by combining (25) and (26). However, we instead rewrite (21) in the following form:[23,41]

$$I_{dsub} = q\mu_n Wn_{so}(x)\frac{dV_F(x)}{dx}\exp\left[-\frac{V_F(x)}{V_{th}}\right], \tag{27}$$

where $V_F(x)$ is the quasi-Fermi potential. We now consider (27) at the device center ($x = 0$), in which case n_{so} can be written in the same form as n_{sm} in (24), except that x_m should be replaced by 0. Here, we invoke the analytical expression for the center potential from (11), which gives

$$\hat{\varphi}_o^{LP} = \left(V_{bi} + V_{FB} - V_{gs} + \frac{V_{ds}}{2}\right)\left[1 - \frac{4}{\pi}\tan^{-1}\left(\sqrt{k}\right)\right]. \tag{28}$$

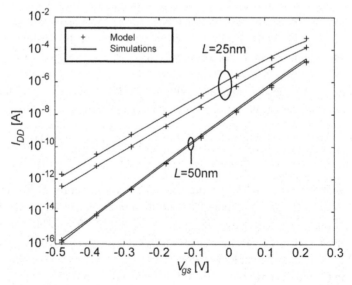

Fig. 12. Comparison of modeled (symbols) and numerically simulated (solid curves) I_{dsub}-V_{gs} subthreshold characteristics for two different gate lengths. The upper curve for each L represents $V_d = 0.5$V and the lower represents $V_{ds}=0.1$V. (After Ref. 40.)

In addition, numerical simulations indicate that both $V_F(x)$ and $dV_F(x)/dx$ are quite independent of V_{gs} at the device center.[41] This assumption typically results in an error of less than 2% in S.

Hence, applying (27) in (26) and using (24) and (28), we find:

$$S = V_{th} \ln(10)\left\{1 - m + \frac{mV_{th}}{2\hat{\phi}_o^{LP}}\left[1 - \frac{2t_{si}}{t_{si} + 2t'_{ox}}\sqrt{\frac{\hat{\phi}_o^{LP}}{\pi V_{th}}}\frac{\exp\left[-\left(\frac{2t_{si}}{t_{si} + 2t'_{ox}}\right)^2\frac{\hat{\phi}_o^{LP}}{V_{th}}\right]}{Erf\left[\frac{t_{si}}{t_{si} + 2t'_{ox}}\sqrt{\frac{\hat{\phi}_o^{LP}}{V_{th}}}\right]}\right]\right\}, \quad (29)$$

where $m = 1 - 4\tan^{-1}\left(\sqrt{k}\right)/\pi$. We note that in deep subthreshold, S reduces to the following simple form[41]

$$S = \frac{\pi \ln(10)V_{th}}{4\tan^{-1}\left(\sqrt{k}\right)}, \quad (30)$$

which approaches 60 mV/decade for long-channel devices ($k \rightarrow 1$).

The above model for the subthreshold slope incorporates the effects of the device dimensions, bias voltages, and source-drain junction doping on the subthreshold slope. Figures 13a and 13b show the scaling of S with channel length and channel thickness, respectively for $V_{ds} = 0.1$ V and $V_{gs} = V_{FB} = -0.481$ V. As expected, we observe that the subthreshold characteristics improve with increasing channel length and reduced body and oxide thicknesses. Thus a long channel or an ultra-thin body device can be optimal for subthreshold applications if the subthreshold slope is a main concern.

2.2.5. *Subthreshold capacitances*

From the above modeling framework, we can also calculate the intrinsic device capacitances in subthreshold. The capacitances are dominated by the inter-electrode coupling, from which analytical expressions for the charge conserving trans- and self-capacitances can readily be derived using conformal mapping techniques. The capacitances are calculated from charges determined by the vertical displacement field on the electrode surfaces using Gauss' law. This field can be derived from the device electrostatics discussed above. Charge conservation requires that the electrode charges associated with the inter-electrode coupling add up to zero total charge.

DG MOSFETs with symmetric biasing can be considered three-terminal devices. Imposing charge conservation [19], such a device will have nine capacitances C_{XY} as indicated in the equivalent circuit in Fig. 14 [20]. Of these, four are independent. The self-capacitances ($X = Y$) and trans-capacitances ($X \neq Y$) reflect the change of the charge on electrode X with a small variation in the voltage on electrode Y.

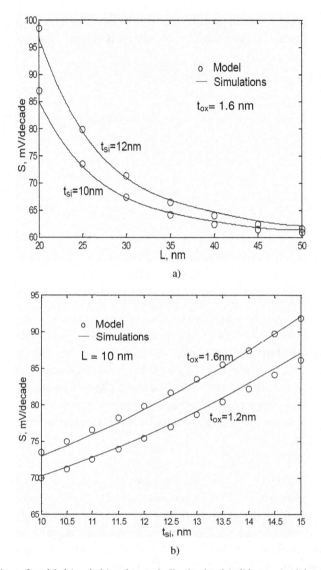

Fig. 13. Comparison of modeled (symbols) and numerically simulated (solid curves) subthreshold slope with a) channel length and b) silicon thickness. V_{ds} =0.1V, V_{gs}= V_{FB} . (After Ref. 41.)

The capacitances are found by taking the derivatives of the electrode charges with respect to the various electrode potentials, i.e., $C_{XY} = \pm dQ_X/dV_Y$, where the plus sign applies when $X = Y$. For the inter-electrode coupling, Q_X is a linear function in V_Y, which means that all the capacitances are bias independent. Owing to symmetry and charge conservation, we find that $C_{GS} = C_{GD} = C_{SG} = C_{DG}$, $C_{DS} = C_{SD}$, $C_{GG} = C_{SG} + C_{DG}$, and $C_{SS} = C_{DD} = (C_{GG} + C_{DS} + C_{SD})/2$.

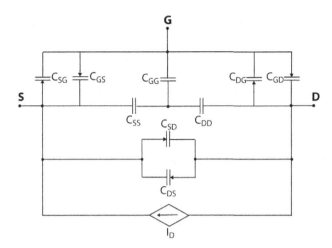

Fig. 14. Equivalent circuit of the DG MOSFET with symmetric gate biasing, indicating a current source and the nine self-and trans-capacitances.

From the DG MOSFET potential distribution in (8), we obtain the perpendicular electric field $E_\perp$ at the electrodes, from which we derive the following analytical expression for the electrode charges associated with the inter-electrode coupling,[25,27]

$$Q_X = \varepsilon_{si} \int_{z_{min}}^{z_{max}} E_\perp dz = i\varepsilon_{si} \int_{u_{min}}^{u_{max}} \left.\frac{\partial \varphi_{DG}^{LP}}{\partial v}\right|_{v \to 0} du$$

$$= \frac{i\varepsilon_{si}}{\pi} \left[V_G \ln\left(\frac{(u-1)(ku+1)}{(u+1)(ku-1)}\right) + V_S \ln\left(\frac{u+1}{ku+1}\right) + V_D \ln\left(\frac{u-1}{ku-1}\right) \right]_{u_{min}}^{u_{max}} . \tag{31}$$

Here, the limits of integration are the appropriate dimensions of the electrodes over which the intrinsic charges are distributed. For the drain and source electrodes, the integration runs from $y = t'_{ox}$ to $t_{si} + t'_{ox}$ for $x = L/2$ and $-L/2$, respectively, or between the corresponding coordinates on the u-axis in the (u,v)-plane. However, to preserve intrinsic total charge neutrality, i.e., $Q_G = Q_S + Q_D$, small regions close to source and drain should be excluded from the integration along the gate electrodes. These are the regions indicated in Fig. 6b between the field line emanating from the source/drain corner (bold curved line) and the part of the boundary corresponding to the oxide gap (vertical dashed line). Since field lines in this region will terminate on the lower side of the source/drain electrode (see dashed curved field lines in Fig. 6b), the associated charges should be excluded and assigned to the extrinsic capacitances.

From the one-corner analysis in Section 2.2.1, we find that the integration associated with the intrinsic gate charge should run between $x = -L/2 + x_0$ and $x = L/2 - x_0$ for the two gates, where $x_0 = 0.339t'_{ox}$, or between the corresponding coordinates along the u-axis in the (u,v)-plane. For the 25 nm long DG device defined above, we obtain the following inter-electrode capacitance values per µm of channel width: $C_{GS} = 0.24$ fF, C_{DS} = 0.005 fF, $C_{GG} = 0.48$ fF.

To check the scaling property of the DG inter-electrode capacitance model, we compared the modeling results with numerical simulations (Atlas) for a range of device lengths between 10 and 35 nm, keeping everything else constant. The results shown in Fig. 15 indicate a good agreement between model and numerical simulations throughout this range. We observe that the source and drain self-capacitances C_{SS} and C_{DD} are not noticeably affected by the change in gate length while the other capacitances reach asymptotic values at $L \approx 30$ nm. This gate length can be interpreted as the lower limit beyond which short-channel effects become important.

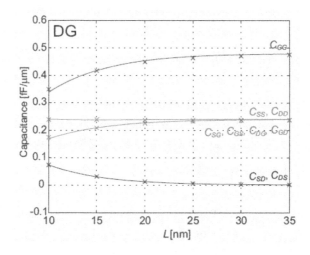

Fig. 15. Modeled intrinsic inter-electrode DG MOSFET capacitances (curves) plotted as a function of device length. Oxide and substrate thicknesses are held constant at $t_{si} = 12$nm and $t_{ox} = 1.6$nm. Numerical simulations (symbols) were carried out for $V_{ds} = 0$ V and $V_{gs} = -0.2$ V. (After Refs. 25, 27.)

2.3. *Self-Consistent Electrostatics at and above Transition in DG MOSFETs*

2.3.1. *Transition voltage*

In Section 2.2.2, we indicated that the inversion charge near source and drain can have a non-negligible influence on the device electrostatics even in deep subthreshold. With increasing gate bias, the electron density will increase throughout the device body and will eventually become an important and even dominant contributor to the body potential. In subthreshold, a small amount of electrons tends to collect in regions of highest potential, i.e., near source and drain and along the S-D axis. With increasing gate bias, a transition eventually takes place where the potential maximum in the G-G direction relocates from the S-D axis to the gates, and the volume inversion is replaced by a surface inversion.

For nanoscale MOSFETs, the value of V_{gs} where this transition takes may conveniently be defined as a transition voltage V_T (other, more operational definitions of the threshold voltage are used in practice). More precisely, V_T is chosen to be the V_{gs} for which the inter-electrode and the inversion charge contributions to the potential exactly add up to the value $V_T - V_{FB}$ at the device center for $V_{ds} = 0$ V. Both these contributions then have a near parabolic and form of equal and opposite amplitude. At this condition, the total potential along the G-G symmetry axis is, therefore, nearly flat and so is the electron concentration. This means that Poisson's equation is reduced to a one-dimension form at device center, allowing the charge density to be expressed in terms of the curvature in the z direction of the total potential, which makes it possible to express V_T in the form of the following Lambert function[23]

$$\frac{V_{bi} - V_T + V_{FB}}{V_{th}} \exp\left(\frac{V_{bi} - V_T + V_{FB}}{V_{th}}\right) = \frac{qn_i^2}{8V_{th}\varepsilon_{si}N_a} \frac{t_{si}\left(t_{si} + 4t_{ox}'\right)}{1 - \frac{4}{\pi}\tan^{-1}\left(\sqrt{k}\right)} \exp\left(\frac{V_{bi}}{V_{th}}\right). \qquad (32)$$

Figure 16 shows a comparison between this model and numerical simulations of V_T versus device length for $V_{ds} = 0$, keeping all other parameters constant. We observe an excellent agreement down to a gate length of about 15 nm, where the assumptions made about zero curvature in the potential in the x-direction at the device center tends to break down. For the DG MOSFET device with $L = 25$ nm considered previously (in Section 2.2), we find the value $V_T \approx 0.25$ V.

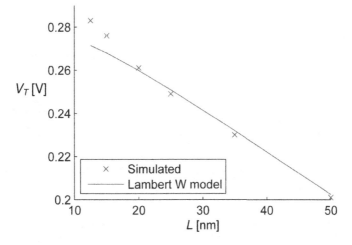

Fig. 16. Scaling of transition voltage with gate length for the DG MOSFET. Results from the model and the numerical simulation are shown as a solid line and as symbols, respectively. All other parameters are taken to be constant. $V_{ds} = 0$ V.(After Ref. 23.)

2.3.2. *Above-transition electrostatics*

In order to derive the device electrostatics within the present modeling framework, we divide the Poisson's equation (1) to (3) into two superimposed parts – the 2D capacitive coupling described by Laplace's equation, and the residual part associated with the charge carriers. The latter then has to be solved self-consistently with the quasi-Fermi potential distribution associated with the drain current. Within the present framework, the modeling of the DG MOSFET proceeds as indicated below and in the flow chart of Fig. 17.

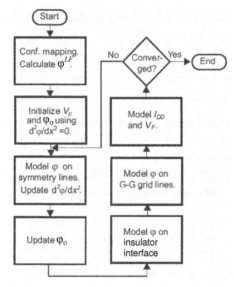

Fig. 17. Flow diagram showing the iteration procedure for self-consistent calculation of the device electrostatics and drain current. (After Refs. 24 and 27.)

Initiation:

- A rough estimate of the quasi-Fermi potential is made – for example, $V_{Fo} = V_{ds}/2$. Also, the total potential φ_o at the device center is estimated – for example, using the corresponding potential in the long-channel limit.[2,3,23]
- Based on this, an approximate potential distribution for the G-G symmetry axis is introduced, such as a parabolic form or the shape derived in the long-channel limit.
- A physics-based modeling expression is established for the potential distribution along the S-D symmetry axis. Its parameters are determined from the boundary conditions at the source and drain (the potential and its first and second derivatives), and at the device center (potential and, for example, its second derivative).
- Using the S-D potential distribution, approximate G-G distributions at different cut-lines can be established, using, for example, the shapes discussed above for the G-G axis.

- Based on the above, the potential at the silicon/insulator interface is modeled.
- The drain current is calculated based on the above initial estimates.

Processing:
- Once the current has been estimated, V_F is up updated
- φ_o is updated by invoking Poisson's equation at the device center, using the model expression for the potential along the S-D axis.
- Based on this, the G-G and S-D potential distributions are updated
- These steps are repeated until a satisfactory accuracy is obtained. Normally, two to four iterations are sufficient.

In strong inversion, the modeling accuracy near the silicon/insulator interface must be carefully evaluated, especially near source and drain. This is done by introducing G-G potential distributions in this region that better reflect the proximity to the contacts, i.e., that the electric field tends to diverge in the corners at $x = \pm L/2$.[27]

Modeled and simulated potential distributions are compared with numerical simulations for a range of bias voltages in Figs. 18 to 20 (same device as in Section 2.2). In all cases, the maximum difference between the modeled potentials and the numerical simulations are less than about 10 millivolts. Figure 18 is an example of the potential distribution along the S-D axis for $V_{gs} = 0.6$ V and $V_{ds} = 0.5$ V, showing the total potential and the contributions from the inter-electrode coupling and from the inversion charge. Figure 19 shows comparisons of potential distributions along the G-G symmetry axis for various gate biases ranging from subthreshold to strong inversion.

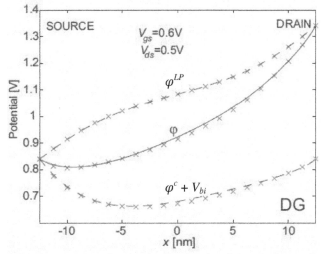

Fig. 18. Modeled (curves) and numerically simulated (symbols) total potential φ and the contributions φ^{LP} and φ^c along the S-D symmetry line for $V_{gs} = 0.6$V and $V_{ds} = 0.5$V. The built-in voltage V_{bi} has been added to φ^c to ease the comparison of the different terms. (After Ref. 27.)

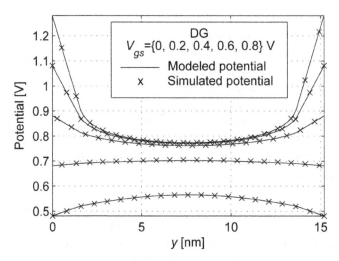

Fig. 19. Modeled and numerically simulated total potential φ along the G-G symmetry axis for $V_{ds} = 0$ V and values of V_{gs} ranging from subthreshold to strong inversion ($V_T = 0.25$ V). (After Ref. 30.)

It is interesting to note that the center potential φ_o tends to converge towards a fixed value with increasing V_{gs} in strong inversion. This behavior will be considered further in conjunction with the unified modeling of multigate devices in Section 4.

In Fig. 20 are presented examples of the strong-inversion distributions of φ and V_F along the silicon/insulator interface of the DG MOSFET for $V_{ds} = 0.6$ V and three different values of gate bias.

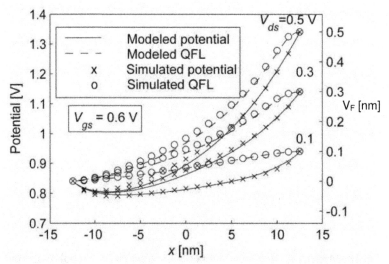

Fig. 20. Modeled and numerically simulated total potential and quasi-Fermi potential along the silicon insulator interface for $V_{gs} = 0.6$ V and different values of V_{ds}. (After Ref. 24.)

2.3.3. *Drain current*

The small dimensions of the devices considered here indicate that the drain current will have the character of both drift-diffusion and ballistic/quasi-ballistic transport. Here, we use a drain current model based on the classical drift-diffusion formalism with constant mobility, which seems to agree quite well with both hydrodynamic and energy transport formalisms.[24] The modeled and simulated drain currents are shown in the I_d-V_{gs} and I_d-V_{DS} characteristics of Figs. 21a and 21b, respectively. Again, an excellent agreement is observed between the modeling and the simulation results within the full range of bias conditions from subthreshold to strong inversion.

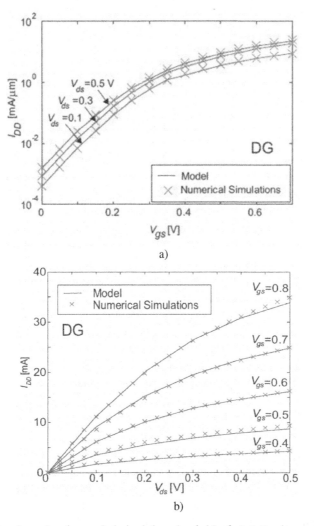

Fig. 21. Modeling (curves) and numerical simulations (symbols) of a) I_d-V_{gs} characteristics and b) I_d-V_{ss} characteristics for a wide range of bias voltages. (After Ref. 24.)

Well above threshold, the inversion charge contribution to the body potential will be dominant, and the electrons tend to screen the effects of inter-electrode capacitive coupling over most of the channel length. Hence, in this limit, the device electrostatics and drain current can be modeled according to a classical long-channel analysis. Thus, the above modeling framework may serve as an excellent starting point for development of more compact modeling expressions suitable for use in circuit simulators. One possibility is to use a set of generic, semi-empirical expressions for the I-V characteristics with parameters that can be extracted to any desired accuracy from this framework.

Typically, such a model may be based on explicit subthreshold and strong-inversion limits that are readily available from the from simple analyses, and on the gate bias dependences of I_d near threshold expressed in terms of parameters that can be extracted from the more comprehensive modeling above. The following is an example of a suitable interpolation function[23,27,43]

$$
\log_{10}(I_d) = \frac{\log_{10}(I_{dsub})}{\left[1 + \left(\frac{\log_{10}(I_{dsub})}{\log_{10}(I_{dinv})}\right)^m\right]^{\frac{1}{m}}}, \tag{33}
$$

where I_{dsub} and I_{dinv} are the subthreshold and strong inversion asymptotes, respectively, and m is a shape parameter that determines the properties of the I_d-V_{gs} characteristics near the transition. In subthreshold, for example, the expression for I_{dsub} in (25) can be used. Another alternative applicable for deep subthreshold is to use the simple expression I_{dsub} = $I_o\exp[(V_{gs} - V_{gso})/S)]$, where I_o is the current calculated at a given value $V_{gs} = V_{gso}$ and the subthreshold slope S is given by (30). The effect of drain-induced barrier lowering (DIBL), and hence the dependence of I_{dsub} on V_{ds}, is implicitly included in I_o. As indicated above, I_{dinv} can be obtained from an analytical long-channel analysis. The interpolation form in (33) is unified in the sense that it uses one continuous expression for all operating regimes. One of the advantages is that it can be differentiated to any order with respect to the bias voltage, which is beneficial in terms of convergence properties when using the model in SPICE-type circuit simulators.

At the outset, the interpolation function in (33) has a free parameter m that can be extracted for each drain bias point from the modeling framework. This extraction should be performed for a given V_{gs} close to the transition voltage. Typically, m is somewhat dependent on V_{ds}, a behavior that can be captured by means of a simple modeling function. For the DG MOSFET considered (same as above), a parabolic least-square fitting routine is suitable. Figure 22 compares the modeled I_d-V_{gs} characteristics of the DG device with numerical simulations. The extracted and modeled m parameter versus V_{ds} is shown in the inset. Since the underlying modeling framework scales with device dimensions and material properties, it also allows the extraction of the scaling properties of the model parameter m.

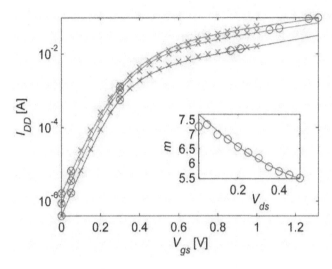

Fig. 22. Modeling (curves) and numerical simulations (symbols) of I_d-V_{gs} characteristics for different V_{ds} using the interpolation function in (33). Circles indicate anchoring points for the modeling. The inset shows the extracted values for m versus V_{ds} (circles) and the parabolic fitting function. (After Ref. 42.)

2.3.4. Above-threshold capacitances

From near threshold to strong inversion, the contribution of the inversion charges will steadily become more significant for the intrinsic device capacitances. From the self-consistent device electrostatics discussed in Section 2.2, we find the perpendicular electric field at the electrodes, from which the total electrode charges Q_S, Q_D, and Q_G and the intrinsic capacitances may be determined using Gauss' law. However, especially Q_S and Q_D may be difficult to determine precisely this way because of strong corner effects. Instead, we concentrate on the terminal mirror charges Q_S^Q, Q_D^Q, and Q_G^Q corresponding to the total body charge Q_B. (The inter-electrode charge contributions Q_S^{LF}, Q_D^{LF}, and Q_G^{LF} were determined separately in Section 2.2.5.) Q_B is obtained by integrating the charge density of the silicon body using the device electrostatics discussed above. We then divide the device into two equal parts separated by the G-G symmetry line and determine the charges Q_{BS} and Q_{BD} associated with the source and drain halves, respectively.

Next, we determine the total gate charge Q_G from the surface field as discussed above and then find the charges Q_{GS} and Q_{GD} associated with the source and drain halves. The two mirror charge contributions at the gate are then obtained as $Q_{GS}^Q = Q_{GS} - Q_{GS}^{LF}$ and $Q_{GD}^Q = Q_{GD} - Q_{GD}^{LF}$, i.e., by subtracting the corresponding inter-electrode charges Q_{GS}^{LF} and Q_{GD}^{LF}. Finally, imposing the condition of overall charge neutrality, the mirror charges at source and drain can be expressed as $Q_S^Q = -\left(Q_{BS} + Q_{GS}^Q\right)$ and $Q_D^Q = -\left(Q_{BD} + Q_{GD}^Q\right)$, from which we obtain the contributions from the body charges on the capacitances.

Figure 23 shows two examples of the modeled capacitances of the DG device versus V_{gs} for $V_{ds} = 0.1$ V and 0.5 V, covering operating conditions from deep subthreshold to strong inversion. Figure 24 shows the modeled capacitances as a function of V_{ds} for V_{gs} = 0.25V and 0.6V. We observe the model compares quite well with the numerical simulations.

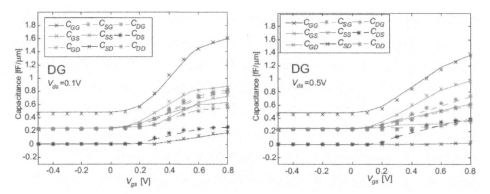

Fig. 23. Modeling (curves) and numerical simulations (symbols) of intrinsic DG MOSFET capacitances versus V_{gs} for V_{ds} = 0.1 V (left) and 0.5 V (right). Threshold voltage: 0.25 V. (After Ref. 27.)

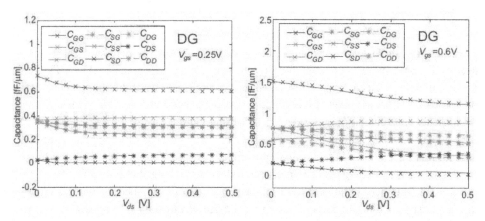

Fig. 24. Modeling (curves) and numerical simulations (symbols) of intrinsic DG MOSFET capacitances versus V_{ds} for V_{gs} = 0.25 V (left) and 0.6 V (right). Threshold voltage: 0.25 V. (After Ref. 27.)

3. Modeling of Circular Gate MOSFETs

3.1. *Subthreshold Electrostatics of GAA MOSFETs Based on 2D Solutions*

So far we have considered the DG MOSFET as a prime candidate for improved device performance in terms of reduced short-channel effect. The improvements stem from the increased gate control. Obviously, still better gate control can be achieved by wrapping the gate around the device body. Examples of such gate-all-around (GAA) MOSFETs are those with circular or square cross sections, denoted CirG and SqG MOSFETs, respectively, shown schematically in Fig. 25.

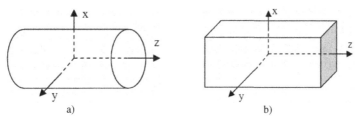

Fig. 25. Schematic view of GAA MOSFETs with circular and square cross sections.

In principle, the derivation of the body potential of GAA MOSFETs requires the solution of the 3D Poisson's equation in (1) or, for the inter-electrode potential, the solution of the 3D Laplace equation in (2). Although the conformal mapping technique is not directly applicable to 3D problems, we may still be able to obtain a very good approximate solution by applying this technique to suitable longitudinal cross-sections (x,z) and equivalent planes, see Fig. 25) of high symmetry GAA MOSFETs, especially those with circular and square cross sections. In fact, we can readily observe significant structural similarities between the 2D inter-electrode potential distributions in such cross-sections and those of DG MOSFETs.

To explore this further, we consider the 3D Laplace equation of (2) for a CirG device of length L and diameter a in Cartesian coordinates. We note that because of the circular symmetry, the x and y axes are equivalent, which means that along the cylinder axis we have $\partial^2 \varphi_{cyl}^{LP} / \partial y^2 = \partial^2 \varphi_{cyl}^{LP} / \partial x^2$ and (2) can be rewritten as,

$$\frac{\partial^2 \varphi_{cyl}^{LP}}{\partial x^2} + \frac{\partial^2 \varphi_{cyl}^{LP}}{\partial \left(\sqrt{2}z\right)^2} = 0. \tag{34}$$

It is interesting to note that this expression also can be interpreted as the 2D Laplace equation for an elongated DG MOSFET of length $\sqrt{2}L$ and thickness a. Alternatively, (34) can be rewritten to correspond to a thinner DG device of thickness $a/\sqrt{2}$ and length L. In any case, the potential distribution obtained for either of these DG devices will, when deformed linearly back to the original 3D cross section, result in a quite satisfactory approximate solution of the potential distribution in any longitudinal ρ-z cross section of the CirG device, and therefore, for the entire extended body of the device. In fact, along the cylinder axis, the 3D Laplace equation will be satisfied everywhere, since this equation is recovered by the operation. The same is true at the source, drain, and gate boundaries. However, a slight error will occur in the off-axis body potential distribution where (34) does not apply precisely. But we have found that this inaccuracy can be compensated quite well by replacing the factor of $\sqrt{2}$ above by $\tau \approx 1.34$.

The logic used for the CirG device can also be applied to the SqG device, specifically when considering the two equivalent longitudinal (x,z) and (y,z) cross sections of Fig. 25b. For the transverse (x,y) cross sections, the modeling of the potential distribution must be based on suitable interpolation expressions. This will be discussed further in Section 3.1.4, where the concept of isomorphic modeling functions is introduced.

Finally, we observe that the above procedure for analyzing the SqG MOSFET can also be generalized to include devices with rectangular (x,y) cross sections (RecG MOSFETs). Assuming that the width in the x and y directions are a and b, respectively, we can now rewrite (34) by observing that $\partial^2 \varphi_{rec}^{LP} / \partial y^2 = \partial^2 \varphi_{rec}^{LP} / \partial (bx/a)^2$ to obtain

$$\frac{\partial^2 \varphi_{rec}^{LP}}{\partial x^2} + \frac{\partial^2 \varphi_{rec}^{LP}}{\partial (\tau_{cyl} z)^2} = 0, \tag{35}$$

where $\tau_{cyl} = \sqrt{1+(a/b)^2}$. However, for cross sections with large aspect ratios b/a, the influence of the top and bottom gates tends to vanish in central x-z cross sections, creating what amounts to a DG MOSFET section in this region. The top and bottom sections of the device then act as two separated halves of a limiting ReG device. For $b/a \gg 1$, the total device behavior approaches that of a vertical DG device or a tall FinFET. These observations will be considered in more detail in Section 4 as part of discussion of unified multigate modeling.

It is also interesting to note that because of the two-fold symmetry of the RecG MOSFET about the (x,z) and (y,z) planes, half of this device becomes a tri-gate MOSFET. Hence, the potential distribution in one half of the former can be applied to the latter. However, the tri-gate device will suffer from fringing field effects that have to be treated separately. But again, a DG solution is a good approximation when dealing with a tall and narrow trigate device, or a FinFET.

Next, as an example, we analyze the electrostatics of the CirG device in more detail based on the results derived in Section 2 for the DG MOSFET.

3.2. Subthreshold Modeling of CirG MOSFETs

As indicated above, the major difference between the DG and the GAA MOSFETs is the strength of the gate control. In the above analysis, this was expressed in terms of the factor $\sqrt{2}$ in (34) and τ in (35). However, the same effect can also be described in terms of so-called characteristic (or natural) lengths[44] λ_{DG} and λ_{cir} for DG and CirG devices, respectively. These parameters are a measure of the subthreshold field penetration depth from source and drain along the S-D symmetry axis. Approximate expressions in terms of device dimension and electrostatic properties can, for example, easily be derived by performing a simplified analysis based on the assumption of parabolic potential distributions in the cross-sections perpendicular to the S-D axis. For the DG and the CirG device, these parameters are given by

$$\lambda_{DG} = \sqrt{\frac{\varepsilon_{si}}{2\varepsilon_{ox}} \left(1 + \frac{\varepsilon_{ox} t_{si}}{4\varepsilon_{si} t_{ox}} \right) t_{si} t_{ox}}, \tag{36}$$

$$\lambda_{cir} = r_{si} \sqrt{\frac{1}{4} + \frac{\varepsilon_{si}}{2\varepsilon_{ox}} \ln\left(1 + \frac{t_{ox}}{r_{si}} \right)}, \tag{37}$$

Respectively,[45,46] where $r_{si} = t_{si}/2$ is the radius of the silicon body in the CirG MOSFET.

Hence, given a CirG device of length L, we calculate the potential distribution of a DG device of the same width but with a length $L\lambda_{DG}/\lambda_{cir}$ and compress it uniformly in the longitudinal direction until we regain the original length L, as indicated in Fig. 26.[23,27] This way, the penetration depth for the resulting 2D distribution in the CirG (x,z) cross section achieves the required value λ_{cir}. This procedure should be equivalent to that described above with $\lambda_{DG}/\lambda_{cir}$ corresponding approximately to $\sqrt{2}$. For the device parameters given in Section 2, we find $\lambda_{DG}/\lambda_{cir} \approx 1.46$, which is within a satisfactory 3.5% of $\sqrt{2}$. But again, we emphasize that this procedure for deriving $\varphi_{cir}^{LP}(x,y)$ does not give an exact solution of the 3D Laplace equation. (Note that the extended body of this device is defined in a similar manner as for the DG MOSFET in Section 2.1.)

The present discussion of the CirG MOSFET shows that we can use the DG MOSFET results of Section 2 as a basis for calculating its subthreshold potential. In doing so, we note that the electrical field at the center of the source and drain electrodes, E_S and E_D, for the CirG device is obtained from the DG expression in (17) by replacing the modulus k by a new modulus k' (see (6)) corresponding to that of the stretched (ρ, z) cross-section to the left in Fig. 26. Also, adjustment for the corner effects and the effect of inversion charge near source may be needed.

Using the device parameters of Section 2 ($L = 25$ nm, $t_{si} = 12$ nm, $t_{ox} = 1.6$ nm), Fig. 27 shows comparisons of subthreshold potential distributions of the CirG MOSFET between the present model and numerical simulations using the Atlas simulator. Figure 27a shows the G-G potential in the radial direction through the silicon body at various position along the device channel for $V_{ds} = V_{gs} = 0$ V, while Fig. 27b shows potential profiles along the S-D symmetry axis for $V_{gs} = 0$ V at different values of V_{ds}. We notice an excellent agreement between the modeling and the numerical simulations, with deviations only in the millivolt range. Comparing the results of the DG MOSFET in Fig. 9 with those of the CirG device in Fig. 27b, we observe a clear reduction of the DIBL effect in the latter, demonstrating the reduced short-channel effects in the GAA structures.

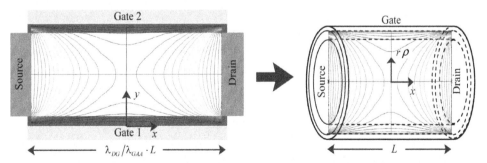

Fig. 26. Schematic illustration of the mapping of a DG MOSFET inter-electrode potential distribution for an extended device of length $L\lambda_{DG}/\lambda_{cir}$ (left) into the longitudinal cross-section of a CirG device of length L (right). After Ref. 27.)

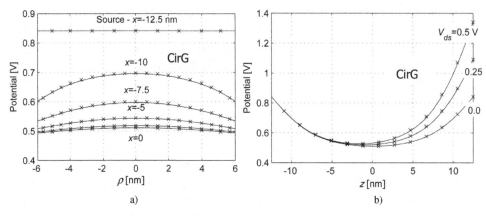

Fig. 27. Modeled and simulated subthreshold potential profiles in the silicon body of a CirG MOSFET along a) the radial direction for $V_{gs} = V_{ds} = 0$ V at different position z, and b) along the S-D axis for $V_{gs} = 0$ V and different V_{ds}. (After Ref. 30.)

To check the scaling properties of the CirG modeling scheme, the potential at the device center was modeled for a range of gate lengths, keeping all other dimensions fixed at the values given in Section 2. As shown in Fig. 28, the modeled values agree very well with numerical simulation down to a gate length of at least 15 nm, demonstrating the very satisfactory scaling properties of the model.

From the above electrostatics, the CirG subthreshold drain current and the intrinsic capacitances can be derived along the same line as for the DG MOSFET in (see Sections 2.2.4 and 2.2.5). Figure 29 shows an example of subthreshold I_{dsub}-V_{ds} characteristics for the above CirG device. The modeling was done by solving the drift-diffusion equation in Section 2.2.4 for the CirG geometry by means of Simpson's formula, and the results compare quite well with numerical simulations. A similar good fit was also obtained using a compact model that was derived along the same line as that for the DG MOSFET in section 2.2.4.

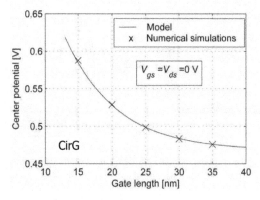

Fig. 28. Modeled and numerically simulated subthreshold center potential versus gate length for CirG MOSFETs. (After Refs. 24 and 30.)

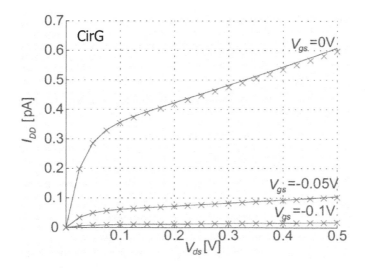

Fig. 29. Comparison of modeled (symbols) and simulated (solid curves) I_{dsub}-V_{ds} subthreshold characteristics for CirG MOSFET. (After Ref. 27.)

Figure 30 shows the modeled scaling of the intrinsic subthreshold CirG capacitances with gate length compared with numerical simulations. We notice that the variation with gate length in this case is much less than was found for the DG MOSFET in Fig. 15, again indicating a significant reduction in the short-channel effects in the CirG device.

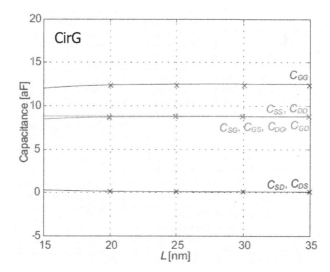

Fig. 30. Modeling (curves) and numerical simulations (symbols) of intrinsic inter-electrode CirG MOSFET capacitances (curves) plotted as a function of device length for $V_{ds} = 0$ V and $V_{gs} = -0.2$ V. (After Ref. 27.)

3.3. *Above-Threshold Modeling of CirG MOSFETs*

The above-threshold modeling of CirG electrostatics, drain current, and capacitances essentially follow the same procedure as that outlined in Section 2.3 for the DG MOSFET. Following the flow diagram of Fig. 17, we arrive, after a few iterations, at a stable, self-consistent solution for the electrostatics, the drain current, and the quasi-Fermi potential distribution.[24,27]

Modeled and simulated potential distributions are compared with numerical simulations for a range of bias voltages in Figs. 31 to 33. In Fig. 31, the potential distribution along the S-D axis is shown for $V_{gs} = 0.6$ V and $V_{ds} = 0.5$ V, indicating the total potential and the contributions from the inter-electrode coupling and from the inversion charge. Figure 32 shows comparisons of potential distributions along the G-G symmetry axis for various gate biases ranging from subthreshold to strong inversion. As for the DG MOSFET, we again observe the saturation of the device center potential φ_o with increasing V_{gs} in strong inversion. In Fig. 33, we present examples of the strong-inversion distributions of φ and V_F along the silicon/insulator interface of the DG MOSFET for $V_{ds} = 0.6$ V and three different values of gate bias.

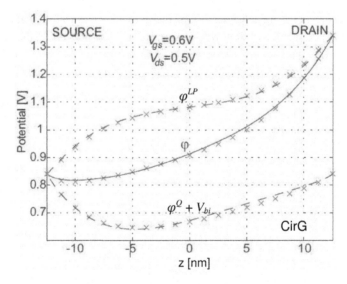

Fig. 31. Modeled (curves) and numerically simulated (symbols) total potential φ and the contributions φ^{LP} and φ^Q along the S-D symmetry line for $V_{gs} = 0.6$V and $V_{ds} = 0.5$V. The built-in voltage V_{bi} has been added to φ^Q to ease the comparison of the different terms. (After Ref. 27.)

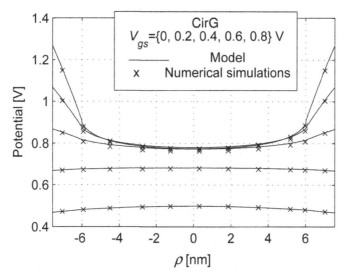

Fig. 32. Modeled and numerically simulated total potential φ along a diameter through the device center for V_{ds} = 0 V and values of V_{gs} ranging from subthreshold to strong inversion. V_T = 0.23 V. (After Ref. 27 and 30.)

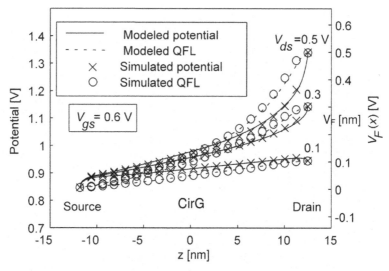

Fig. 33. Modeled and numerically simulated total potential and quasi-Fermi potential along the silicon insulator interface for V_{gs} = 0.6 V at different values of V_{ds}. (After Refs. 24 and 27.)

The corresponding modeled and simulated drain currents are shown in the I_d-V_{gs} and I_d-V_{DS} characteristics of Figs. 34a and 34b, respectively. Again, an excellent agreement is observed between the modeling and the simulation results within the full range of bias conditions from subthreshold to strong inversion. Comparing these results with those of the DG MOSFET in Fig. 20, we notice a diminished dependence of drain bias in the CirG subthreshold current, indicating a significantly reduced DIBL effect.

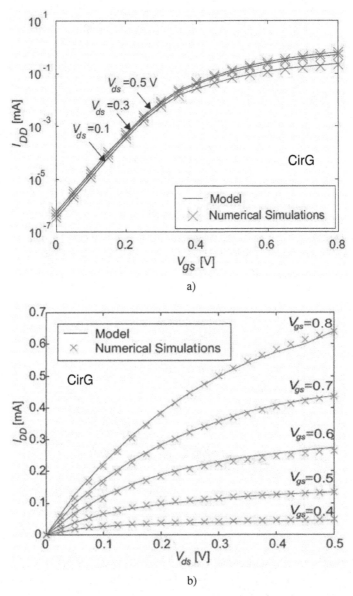

a)

b)

Fig. 34. Modeling (curves) and numerical simulations (symbols) of a) I_d-V_{gs} characteristics and b) I_d-V_{ds} characteristics of CirG MOSFET for a wide range of bias voltages. (After Refs. 24 and 27.)

Finally, we present examples of modeled capacitances of the CirG device covering operating conditions from deep subthreshold to strong inversion. The modeling was performed along the same line as for the DG MOSFET in Section 2.3.4. In Fig. 35, is shown the dependence of capacitances on V_{gs} for $V_{ds} = 0.1$ V and 0.5 V, and in Fig. 36 dependence on V_{ds} for $V_{gs} = 0.25$V and 0.6V is shown. We again observe that the model compares quite well with the numerical simulations.

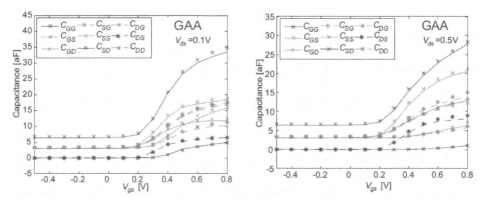

Fig. 35. Modeling (curves) and numerical simulations (symbols) of intrinsic CirG MOSFET capacitances versus V_{gs} for $V_{ds} = 0.1$ V (left) and 0.5 V (right). $V_T = 0.23$ V. (After Refs. 24 and 27.)

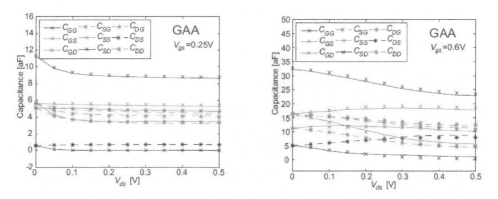

Fig. 36. Modeling (curves) and numerical simulations (symbols) of intrinsic CirG MOSFET capacitances versus V_{ds} for $V_{gs} = 0.25$ V (left) and 0.6 V (right). $V_T = 0.23$ V. (After Ref. 27.)

4. Unified Analytical Modeling of MugFETs

The above results obtained for DG and CirG MOSFETs using conformal mapping techniques can also be applied for the purpose of deriving simpler, unified models for these devices as well as for other GAA MOSFETs such as SqG and RecG devices, and even trigates and FinFETs. Such modeling can be based on the symmetry properties of the MugFET devices and the use of isomorphic modeling functions in the directions perpendicular to the gates in the central regions of the devices. From this, an approximate, analytic, long-channel solution for inter-electrode body potential distribution can obtained from the Laplace equation. However in order to precisely account for short-channel effects, the electrostatics near source and drain needs special attention. We observe that sufficiently close to these electrodes, the potential will be governed by a 1D Laplace equation, which indicates that the potential is reasonably constant in directions perpendicular to the channel axis. For the DG MOSFET, the geometric range of validity of this 1D approximation can be visualized as an equilateral triangle with the base covering the surface of source or drain and with a height corresponding to the characteristic length λ_{dg}. For the CirG and the SqG devices, the range of validity would be a right circular cone and a regular pyramid, respectively, with heights corresponding to the appropriate characteristic lengths. This allows us to introduce alternate functional forms by applying auxiliary, analytical boundary conditions obtained from the conformal mapping analysis discussed in Section 2.

The devices considered here have the same specifications as used for the DG and the CirG in Sections 1 to 3, except that now we assume $L = 30$ nm and $t_{ox} = 1.5$ nm. For the present modeling, the spatial coordinates (x,y,z) are now referred to a Cartesian coordinate system with its origin at the center of the source electrode and the z axis is along the S-D symmetry axis. For the SqG MOSFET, the x and y axes are perpendicular to the gate electrodes while $\rho = \sqrt{x^2 + y^2}$ is the radial coordinate in the CirG device (see Fig. 37).

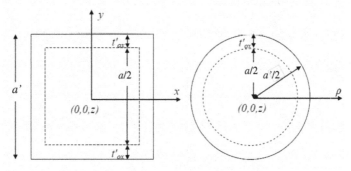

Fig. 37. Transverse cross sections of SqG and CirG MOSFET.

4.1. *Isomorphic Modeling of CirG and SqG MOSFETs in Subthreshold*

4.1.1. *A simple long-channel model*

We start by considering a simple model for long-channel devices where 1D phenomena near source and drain are neglected and where a parabolic distribution of the potential distribution perpendicular to the gates is assumed. Since the extended body/gate interface is $V_{gs} - V_{FB}$, we write the body potential as

$$\varphi(x, y, z) = \hat{\varphi}(x, y, z) + V_{gs} - V_{FB} \tag{38}$$

where $\hat{\varphi}(x, y, z)$ is the potential referred to this interface. Hence, using the proper boundary condition at the gate, we can model the body potential of the CirG and SqG MOSFETs using the following simple, isomorphic modeling functions:[32,33]

$$\hat{\phi}(x, y, z) = \hat{\phi}(0,0,z) \begin{cases} \left[1 - \left(\dfrac{2\rho}{a'}\right)^2\right] & CirG \\[2em] \left[1 - \left(\dfrac{2x}{a'}\right)^2\right]\left[1 - \left(\dfrac{2y}{a'}\right)^2\right] & SqG \end{cases} \tag{39}$$

Here, the parameter *a'* is the thickness of the extended body that includes the gate insulator region, i.e., $a' = a + 2t'_{ox}$. We note that (39) obeys the boundary conditions exactly at the extended body/gate interface.

Applying (39) to the 3D Laplace equation, we readily obtain the following expression for the potential distribution along the z-axis for both devices

$$\hat{\phi}(0,0,z) = \left[\left(V_{bi} - V_{gs} + V_{FB}\right)\sinh\left(\frac{L-z}{\lambda}\right) + \left(V_{bi} + V_{ds} - V_{gs} + V_{FB}\right)\sinh\left(\frac{z}{\lambda}\right)\right]\Big/\sinh\left(\frac{L}{\lambda}\right), \tag{40}$$

where the characteristic (natural) lengths of field penetration from source-drain into the device body is $\lambda = \lambda_{cir} = \lambda_{sq} = a'/4$ for the CirG and SqG devices. Combined with (39), this gives approximate 2D and 3D potential distributions for the respective long-channel body.

4.1.2. *Short-channel modeling of CirG and SqG devices in subthreshold*

For application in short-channel devices, the above model can be improved as follows: The model potential for the (*x*,*y*) cross sections should properly reflect the flattening of the potential inside the silicon body near source and drain, and (39) should contain higher order terms in the transverse coordinates beyond the parabolic approximation. These adjustments are also tied to the potential distribution along the z-axis.[32,33]

In practice, the potential distribution $\hat{\phi}(0,0,z)$ can readily be improved by introducing auxiliary boundary conditions derived from the conformal mapping analysis discussed in Sections 2 and 3. Two such boundary conditions are the center potential φ_o and the electrical field E_S at the face center of the source (see Sections 2.2 and 3.2). For the CirG and SqG devices, these extra boundary conditions can easily be satisfied in (40) by introducing a z-dependent λ of the form: [32,33]

$$\lambda(z) = \lambda_c + \eta(z - L/2)^2, \tag{41}$$

where λ_c and η are parameters derived by combining (39) with φ_o from (11) and E_S from (17) as follows:

$$\lambda_c = \frac{L/2}{\cosh^{-1}\left(\left[1 - \dfrac{4}{\pi}\tan^{-1}\left(\sqrt{k'}\right)\right]^{-1}\right)}, \tag{42}$$

$$\eta \approx \left(\frac{2}{L}\right)^2 \left(\frac{V_{bi} - V_{gs} + V_{FB}}{E_s} - \lambda_c\right) \approx \frac{2}{L}\left[\frac{\pi}{4}\frac{(1+k')/\sqrt{k'}}{K\left(2\sqrt{k'}/(1+k')\right)} - \frac{2\lambda_c}{L}\right]. \tag{43}$$

Note that here the modulus k of the DG device is replaced by k' for the GAA MOSFET. k' is obtained from (6) by multiplying the right-hand side by $2/\tau$ (alternatively, see Section 3.2). For devices of practical interest, k' is typically close to 1, which has allowed us to omit a small V_{ds} dependent term in η.

An updated model for $\hat{\varphi}(0,0,z)$ is obtained by applying (41) – (43) in (40). Combined with (38), this gives a quite precise description of $\varphi(0,0,z)$ for the entire length of the S-D axis. Figure 38 shows model calculations of $\varphi(0,0,z)$ for the CirG and the SqG devices. In both cases, we observe an excellent agreement with numerical simulations (Atlas). We also observe that the values of the potentials obtained for the two GAA devices are very similar.

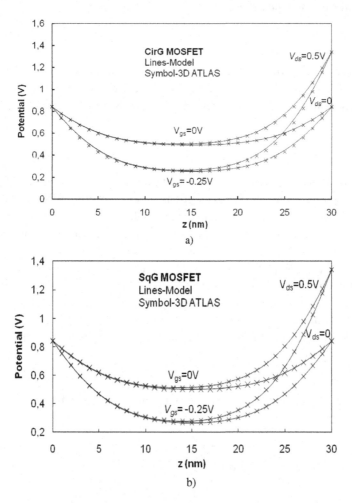

Fig. 38. Model calculations of potential distribution (curves) along a) the D-S axis for the CirG MOSFET and b) for the SqG MOSFET for different values of V_{ds} and V_{gs}. The results are compared with numerical simulations (symbols). (After Ref. 32 (SqG) and Ref. 33 (CirG).)

Next, the modeled potential distributions of the (x,y) plane of (39) are adjusted as follows by adding higher order terms in the isomorphic modeling functions. For the devices considered, adding the next order term suffices in subthreshold,[32,33] giving,

$$\hat{\varphi}(x,y,z)=\hat{\varphi}(0,0,z)\begin{cases}\alpha_{cir}\left[1-\left(\dfrac{2\rho}{a'}\right)^2\right]+\beta_{cir}\left[1-\left(\dfrac{2\rho}{a'}\right)^4\right] & CirG \\[3mm] \alpha_{sq}\left[1-\left(\dfrac{2x}{a'}\right)^2\right]\left[1-\left(\dfrac{2y}{a'}\right)^2\right]+\beta_{sq}\left[1-\left(\dfrac{2x}{a'}\right)^4\right]\left[1-\left(\dfrac{2y}{a'}\right)^4\right] & SqG\end{cases}. \quad (44)$$

Here, the parameters α and β are determined by two conditions: i) The potential at the z-axis ($x = y = \rho = 0$) requires that $\alpha + \beta = 1$; ii) the Laplace equation requires that the curvature of the potential along the z-axis is the negative of the sum of the curvatures in orthogonal transverse directions. From the latter we find for the device center[33]

$$\alpha = \left(\frac{a'}{4\lambda_c}\right)^2\frac{V_{bi}-V_{gs}+V_{FB}+V_{ds}/2}{\hat{\varphi}_o\cosh\left(L/2\lambda_c\right)}\left[1+\eta L\tanh\left(L/2\lambda_c\right)\right]=\left(\frac{a'}{4\lambda_c}\right)^2\left[1+\eta L\tanh\left(L/2\lambda_c\right)\right]. \quad (45)$$

For the CirG and SqG devices considered, we obtain the parameter values $\alpha_{cir} = 1.32$, $\beta_{cir} = -0.32$ and $\alpha_{sq} = 1.31$, $\beta_{sq} = -0.31$, respectively.

We note that for added precision, even higher order terms may be added in the model expressions for the potential in (44), but at the expense of manageability.

Figure 39 shows the application of the updated subthreshold model for the potential in the high-symmetry directions of the (x,y) plane of the SqG device, and in the radial direction of the CirG device at $z = L/2$. In all cases, we observe an excellent agreement with numerical simulations. We find a potential difference of about 6 mV between the SqG and CirG devices at the device center. The slightly lower potential for the CirG MOSFET is consistent with the expected better gate control for this device.

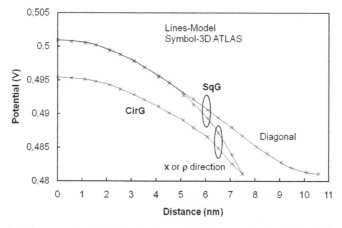

Fig. 39. Model calculations of potential distribution (curves) in the central, extended (x,y) plane in the radial directions of the CirG device, and in high-symmetry directions in the SqG device. The results are compared with numerical simulations (symbols). $V_{ds} = V_{gs} = 0$ V. (After Ref. 33.)

A remaining task is to account for the flattening of the potential distribution of the (*x*,*y*) planes inside the silicon body close to source and drain. Typically, an adjustment of the potential distribution is warranted within the characteristic distance of these electrodes, now defined by $\lambda_s = \lambda(0) = \lambda_c + \eta(L/2)^2$. However, in the region of the central device, given by $\lambda_s \leq z \leq L - \lambda_s$, we retain the potential profiles according to the updated expressions above. This gives rise to the potential distributions in the central (*x*,*y*) plane of the SqG and CirG devices shown in Fig. 39 and the corresponding 2D equipotential contours indicated schematically in the upper right of Fig. 40.

For the regions near source/drain, $z < \lambda_s$ and $z > L - \lambda_s$, we propose the introduction of *z*-dependent equipotential areas in $\varphi(x,y,z)$ in a region about the *z*-axis.[32,33] As indicated above, the range of validity of this approximation would be within a right circular cone for the CirG device and a regular pyramid for the SqG device, both with its base at source and drain and with a height corresponding to the characteristic penetration depth λ_s. This range is indicated as triangular 2D projections in the lower part of Fig. 40. The vertical line through the triangle at the source side represents an equipotential cut, which is shown as a gray region in the (*x*,*y*) cross sections to the upper left. Hence, the width $a_o(z)$ of the equipotential area is zero at $z = \lambda_s$ and $z = L - \lambda_s$, and is assumed to increase linearly towards the source or drain to cover these contacts at $z = 0$ and $z = L$, respectively, i.e.,

$$a_o = \begin{cases} a(1-z/\lambda_s) & \text{for } z < \lambda_s \\ 0 & \text{for } \lambda_s \leq z \leq L-\lambda_s \\ a(1-(L-z)/\lambda_s) & \text{for } z > L-\lambda_s \end{cases} \qquad (46)$$

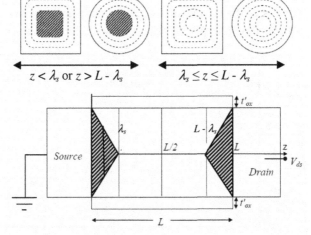

Fig. 40. Longitudinal cross section of the SqG and CirG devices along the channel axis (lower center) with schematic equipotential contour plots in (*x*,*y*) planes near source/drain (upper left) and near the middle (upper right). (After Ref. 33.)

Beyond the equipotential areas, we use (x,y) potential profiles based on the above, but suitably scaled such that they match the equipotential areas at their periphery (exactly for the CirG device and reasonably close for the SqG device).

Hence, for the CirG and SqG devices, the modified potential distribution for $z < \lambda_s$ and $z > L - \lambda_s$ can be expressed as follows inside the equipotential area

$$\varphi(x, y, z) = \varphi(0,0,z) = V_{gs} - V_{FB} + \hat{\varphi}(0,0,z), \tag{47}$$

and outside the dark area as

$$\varphi(x, y, z) = V_{gs} - V_{FB} +$$

$$\hat{\varphi}(0,0,z)\left\{ \begin{array}{cc} \dfrac{\alpha_{cir}\left[1-\left(\dfrac{2\rho}{a'}\right)^2\right]+\beta_{cir}\left[1-\left(\dfrac{2\rho}{a'}\right)^4\right]}{\alpha_{cir}\left[1-\left(\dfrac{a_0(z)}{a'}\right)^2\right]+\beta_{cir}\left[1-\left(\dfrac{a_0(z)}{a'}\right)^4\right]} & CirG \\[30pt] \dfrac{\alpha_{sq}\left[1-\left(\dfrac{2x}{a'}\right)^2\right]\left[1-\left(\dfrac{2y}{a'}\right)^2\right]+\beta_{sq}\left[1-\left(\dfrac{2x}{a'}\right)^4\right]\left[1-\left(\dfrac{2y}{a'}\right)^4\right]}{\alpha_{sq}\left[1-\left(\dfrac{a_0(z)}{a'}\right)^2\right]+\beta_{sq}\left[1-\left(\dfrac{a_0(z)}{a'}\right)^4\right]} & SqG \end{array} \right. \tag{48}$$

The above modeling results are illustrated in Fig. 41, where the potential distribution within the silicon body along the radial direction of the CirG device, and along the x direction in the SqG device are compared with numerical simulations for different positions z along the channel axis. Note that the potential profile at $z = L/8$ is within the near-field range of the source where $z < \lambda_s$, indicated by the flat portion of the modeled profiles for distances less than about 2 nm from the S-D axis.

Finally, the scaling properties of this model are investigated by comparing the center potential φ_o versus gate length of these devices using numerical simulations. The results shown in Fig. 42 indicate a very good match between the model and the simulations for gate lengths down to about 10 nm for all devices. Below this, we may expect increasing mismatch both related to the approximations made and because of quantum mechanical effects, which are not considered here.

Analytical unified modeling of the subthreshold drain current in CirG and SqG MOSFETs can be performed along the same lines as discussed in Section 2.2.4 and 3.2, by assuming a normal distribution of inversion charge about the barrier maximum. Figures 43a and 43b show comparisons of modeled and simulated I_d-V_{gs} characteristics for the CirG and SqG devices, each for two values of V_{ds}. Note that for the CirG device, the characteristics have been extended to the strong inversion regime using the simplified interpolation expression of (33) in combination with strong inversion limits.

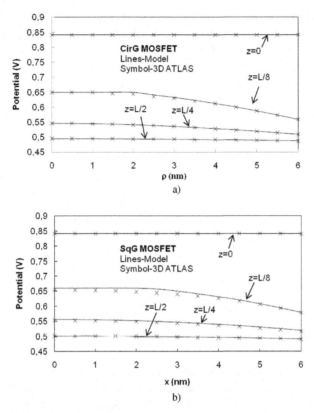

a)

b)

Fig. 41. Model calculations of potential distribution (curves) at different positions along the z direction for a) the radial direction of the CirG device, and b) the x direction in the SqG device. The results are compared with numerical simulations (symbols). $V_{ds} = V_{gs} = 0$ V. (After Ref. 33 (CirG) and Ref. 32 (SqG).)

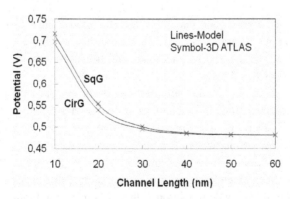

Fig. 42. Comparison of model and numerical simulations of the center potential φ_o versus channel length for the CirG and SqG MOSFET. $V_{ds} = V_{gs} = 0$ V. (After Ref. 33.)

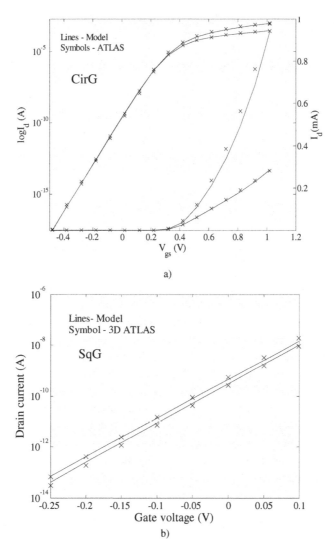

Fig. 43. Comparison of modeled and numerically simulated I_d-V_{gs} characteristics for a) the CirG MOSFET with V_{ds} = 0.1 V and 0.5 V (subthreshold to strong inversion), and b) the SqG MOSFET with V_{ds} = 0.1 V and 1 V (subthreshold). (After Ref. 33 (CirG) and Ref. 32 (SqG).)

Based on the above modeling, we can readily obtain analytical expressions for the subthreshold capacitances of the CirG and SqG devices. In the short-channel case, we derive the following expressions for the perpendicular electric field at the interface between the extended body and the gate in the CirG device:

$$E_{G\perp cir}(z) = -\frac{\partial \hat{\varphi}(\rho, z)}{\partial \rho}\bigg|_{\rho = a'/2} = \frac{4}{a}\hat{\varphi}(0,0,z)(1+\beta_{cir})$$

$$\times \begin{cases} 1, & \lambda_s \le z \le L-\lambda_s \\ \left\{\alpha_{cir}\left[1-\left(\frac{a_0(z)}{a'}\right)^2\right] + \beta_{cir}\left[1-\left(\frac{a_0(z)}{a'}\right)^4\right]\right\}^{-1}, & z < \lambda_s \text{ and } z > L-\lambda_s \end{cases} \tag{49}$$

Hence, the charge per unit length of the channel becomes $q_{gcir}(z) = -\pi a' \varepsilon_{si} E_{g\perp cir}(z)$, where W is the device width, and the total gate charge is obtained from (49) in combination with the modeling results above,

$$\begin{aligned} Q_{Gcir} &= -\pi a'(1+\beta_{cir})\varepsilon_{si}\int_0^L E_{G\perp cir}(0,0,z)dz \\ &\approx -\frac{4\pi}{e}(\lambda_c + \sigma\lambda_s)(1+\beta_{cir})\varepsilon_{si}\left(2V_{bi} + V_s + V_d - 2V_g - 2V_{FB}\right), \end{aligned} \tag{50}$$

where V_s, V_d, and V_g are the source, drain and gate voltages, respectively, and $e = 2.718\ldots$ In (50), the term $\sigma\lambda_s$ in the first parentheses of the final expression represents the fraction of the gate charge associated with the regions near source ($z < \lambda_s$) and drain ($z > L - \lambda_s$), and the term λ_c comes from the central part of the gate region ($\lambda_s \le z \le L - \lambda_s$). Also, we have used the condition $\alpha_{cir} + \beta_{cir} = 1$, made the approximations $\lambda(z) \approx \lambda_c$ in the central gate region, $\lambda(z) \approx \lambda_s$ near source and drain, and retained only the leading terms in the hyperbolic functions of $\hat{\varphi}(0,0,z)$ in (40) when appropriate. For the device considered, we find that $\sigma \approx 6.1$, indicating that the major contribution to the gate charge comes from within the field penetration depth λ_s of the source and drain, as expected.

From the definitions of the self- and trans-capacitances, invoking symmetry relation and charge conservation, we obtain from (50) the following approximate subthreshold results for the CirG device:

$$C_{GGir} = \frac{dQ_{Gcir}}{dV_g} \approx \frac{8\pi}{e}(\lambda_c + \sigma\lambda_s)(1+\beta_{cir})\varepsilon_{si}, \tag{51}$$

$$C_{GScir} = C_{GDcir} = C_{SGcir} = C_{DGcir} = C_{GGcir}/2, \tag{52}$$

$$C_{SScir} = C_{DDcir} = (C_{GGcir} + C_{DScir} + C_{SDcir})/2 \approx C_{GGcir}/2. \tag{53}$$

The approximation in (53), i.e., $C_{DScir} = C_{SDcir} << C_{GGcir}$, is assumed to be applicable when L exceeds about $2\lambda_c$, which is typically the case for well-designed devices.

A similar approach can be used for the SqG device based on the appropriate potential distributions in the unified model above. We note that for any given z coordinate the magnitude of the vertical electrical field $E_{g\perp cir}$ is constant at the circular gate, while it varies with the position on the gate of the SqG device. However, performing the integration of $E_{g\perp sq}$ at z over the four identical gate sidewalls gives the gate charge per unit area. The subsequent integration over z results in the total gate charge. From this, the resulting gate self-capacitance for the SqG devices becomes:

$$C_{GGsq} = \frac{dQ_{GSq}}{dV_g} \approx \frac{32}{15e}\left(10+14\beta_{sq}\right)\left(\lambda_c + \sigma\lambda_s\right)\varepsilon_{si}\varepsilon_o. \tag{54}$$

The relationships between the self- and transcapacitances of (52) and (53) also apply to this device.

From (54), we have extracted modeled values of C_{GGsq} and obtained C_{GSsq} using (52). These results compare quite well with the corresponding numerically simulated values as shown in Table 3.

Table 3. Modeled SqG capacitances compared with numerical simulations.

CirG	Model	Atlas
C_{GSsq}	$4.04\text{x}10^{-18}$ F	$4.11\text{x}10^{-18}$ F
C_{GGsq}	$8.09\text{x}10^{-18}$ F	$8.23\text{x}10^{-18}$ F

4.1.3. *Rectangular gate and trigate MOSFETs*

As briefly discussed in Section 3.1, the procedure used for analyzing the SqG MOSFET can be extended to include rectangular gate (RecG) devices. Based on this, the above modeling technique can be generalized to include DG, trigate, and FinFET devices, laying the groundwork for a unified, compact modeling framework for multigate MOSFETs.

For the RecG device, we rewrite the isomorphic expression in (44) to make it suitable for a rectangular cross section (x,y) of silicon widths a and b (see Fig. 44),

$$\hat{\varphi}(x,y,z) = \hat{\varphi}(0,0,z)\left\{\alpha_{rec}\left[1-\left(\frac{2x}{a'}\right)^2\right]\left[1-\left(\frac{2y}{b'}\right)^2\right] + \beta_{rec}\left[1-\left(\frac{2x}{a'}\right)^4\right]\left[1-\left(\frac{2y}{b'}\right)^4\right]\right\}, \tag{55}$$

were $a' = a + 2t'_{ox}$, $b' = b + 2t'_{ox}$. Again $\hat{\varphi}(0,0,z)$ is given by (40) in combination with (41) to (43), α_{rec} is given by (45) when replacing $a'/4$ by $[8(a'^{-2} + b'^{-2})]^{-1/2}$, and $\beta_{rec} = 1 - \alpha_{rec}$. But, as for the SqG device, corrections are also required in the (x,y) planes within the characteristic length λ_s of the source and drain (see section 4.1.2). Here the potential distribution of the RecG device is still described by (48), except that $2y/a'$ should be replaced by $2y/b'$ everywhere and the denominator in this expression may be rewritten as follows:

$$\alpha_{rec}\left[1-\left(\frac{a_0(z)}{a'}\right)^2\cos^2(\vartheta)-\left(\frac{b_0(z)}{b'}\right)^2\sin^2(\vartheta)\right]+\beta_{rec}\left[1-\left(\frac{a_0(z)}{a'}\right)^4\cos^2(\vartheta)-\left(\frac{b_0(z)}{b'}\right)^4\sin^2(\vartheta)\right]$$

Here $a_0(z) = a(1-z/\lambda_s)$, $b_0(z) = b(1-z/\lambda_s)$, and ϑ is the angle relative to the x direction in the (x,y) plane.

Based on this, we obtain the expression for the potential throughout the device body. This is verified in Fig. 45, which shows $\varphi(0,0,z)$ compared with the numerical simulations for a device with aspect ratio $A_R = b/a = 2$. However, we observe note that (55) will only be directly valid for a limited range of aspect ratios. The reason is that, with increasing A_R, the effects of the top and bottom gates on the potential will diminish and eventually disappear, first in the central (x,z) plane (at $y = 0$), and then spread vertically creating a growing vertical section of the device above and below $y = 0$.

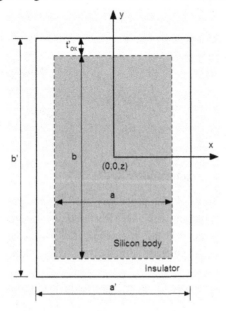

Fig. 44. The (x,y) cross section of a RecG MOSFET with aspect ratio $A_R = b/a = 2$.

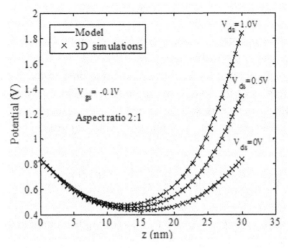

Fig. 45. Modeled potential in RecG MOSFET along the z axis compared to numerical simulations for $V_{gs} = -0.1$ V and different drain biases. $L = 30$ nm, $a = 12$ nm, $b = 24$ nm, $t_{ox} = 1.5$ nm, $\varepsilon_{ox} = 7$. (After Ref. 47.)

This central section will then behave as a vertical DG device of thickness a (see inset in Fig. 46). The onset of this behavior is indicated when the center potential of the RecG device attains the same value as that of a DG MOSFET of the same thickness a, which takes place at about $A_R \approx 4$.[47] We observe that when A_R increases beyond this value, the height of the central DG region increases linearly with A_R while the potential distribution near the top and bottom gates remain invariant with the shape found for $A_R = 4$. Hence, the present modeling permits aspect scaling of rectangular MOSFETs over the full range from $A_R = 1$ (square gate) to infinity, which includes all FinFETs and DG devices. Figure 46 shows a comparison of the modeled potential compared with the numerical simulations along the y axis for devices with $A_R = 4$ and $A_R = 5$.

Moreover, we note that because of the two-fold symmetry of the RecG MOSFETs about the (x,z) plane, half of this device becomes a trigate MOSFET. Hence, the potential distribution in a symmetric half of the former can be applied to the latter, as illustrated by the equipotential contours in the (x,y) plane of a RecG device in the inset of Fig. 47. Hence, a complete subthreshold model of the tri-gate device can also then be formulated along the same line as of the RecG MOSFET. However, the lack of a gate electrode at the lower boundary of the trigate device gives rise to fringing fields, mostly close to the source/drain contacts. As the potential solution near the top of the barrier is not significantly affected by these fringing fields, the solution in that region can still be used to calculate the drain current.

A trigate device with a relatively large aspect ratio is needed for high current drive capability and for reducing the relative effects of the fringing fields. Such a device corresponds to a FinFET. The same is true for RecG devices with a large A_R. Figure 47 shows the comparison of modeled potential compared with the numerical simulations for trigate device with a silicon height of 30 nm and width of 12 nm, corresponding to a

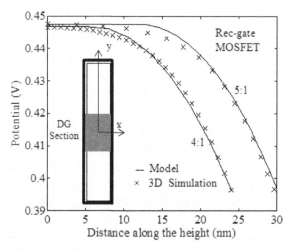

Fig. 46. Modeled potential compared to numerical simulations along the height (y) direction of RecG devices with $A_R = 4$ and 5, $V_{ds} = 0$ V, $V_{gs} = -0.1$V. The inset illustrates schematically an (x,y) cross section of a device with a DG section when $A_R > 4$. (After Ref. 47.)

RecG device with $A_R = 5$. The inset show schematically the equipotential contours of a corresponding pair of RecG and trigate devices.

Modeling of the subthreshold drain current in the assortment of devices discussed here can be performed along the same lines as discussed in Sections 2.2.4, 3.2, and 4.1.2. Figure 48 shows comparisons of modeled and simulated subthreshold I_d-V_{gs} characteristics for a RecG device with $\kappa = 2$ and a trigate device derived from a RecG device with $\kappa = 5$, each for two values of V_{ds}.

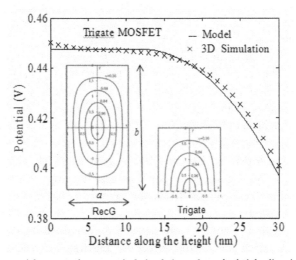

Fig. 47. Modeled potential compared to numerical simulations along the height direction for a trigate device with height 30 nm and width 12 nm, corresponding to a RecG device with $\kappa = 5$. $V_{ds} = 0$ V, $V_{gs} = -0.1$V. The inset shows schematically the equipotential contours in (x,y) cross sections of a RecG MOSFET (left) and the corresponding trigate device (right). (After Ref. 47.)

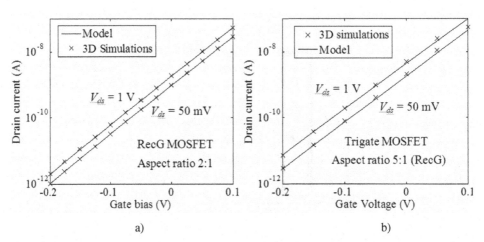

a)   b)

Fig. 48. Modeled subthreshold current in a) RecG device with $A_R = 2$ and b) a trigate device derived from a RecG device with $A_R = 5$, both compared with the numerical simulations. (After Ref. 47.)

4.2. *Modeling of GAA MOSFETs in Strong Inversion*

The unified modeling of subthreshold electrostatics in GAA MOSFETs in Section 4.1 was based on the use of compact isomorphic modeling in combination with a power expansion of the potential in the (x,y) plane. This strongly reflects the symmetry of the cross-sections considered (circular, square, rectangular, etc). For strong inversion, we adopt a similar isomorphic modeling approach, but instead of a power expansion of the cross sectional potentials, we instead introduce functional forms for the (x,y) plane based on rigorous solutions for long-channel DG devices. In this case, the potential distribution along the S-D symmetry axis will also reflect the strong effects of the of inversion charge, that tend to screen the influence of the source and drain from the interior of the device. Based on this analysis, we can derive analytical expressions for the drain current and for the intrinsic capacitances. Here, we emphasize a detailed modeling of the SqG MOSFETs, but a similar approach can be applied to other multigate devices, such as was done for subthreshold in Section 4.1.3. As background for this analysis, we first consider the modeling of DG MOSFETs in strong inversion.

4.2.1. *Strong inversion electrostatics in DG MOSFETs*

We start the strong inversion analysis of the DG MOSFET by considering the electrostatics along the gate-to-gate direction (x direction) in regions of the channel where we can assume that $\partial^2 \varphi(0,0,z)/\partial z^2 = 0$ for $V_{ds} = 0$. For a lightly doped device, the potential φ_o at the S-D axis can be written in terms of Poisson's equation as,

$$\frac{\partial^2 \varphi(x,z)}{\partial x^2} = \frac{q}{\varepsilon_{si}} \frac{n_i^2}{N_a} \exp\left(\frac{\varphi_o}{V_{th}}\right). \tag{56}$$

We note that (56) has the same form as Poisson's equation along the S-D axis of the undoped, long-channel DG MOSFET. The only difference is that n_i has been replaced by n_i^2/N_a. Hence, the DG potential along the x direction is obtained simply by doing the same replacement of n_i in the corresponding, analytical solution for the undoped DG device potential by Taur,[2] i,e.,

$$\varphi(x) = \varphi_o - 2V_{th} \ln\left[\cos(hx)\right]. \tag{57}$$

where

$$h = \sqrt{\frac{qn_i^2 \exp(\varphi_o/V_{th})}{2\varepsilon_{si}V_{th}N_a}} = \frac{\pi}{a} \exp\left(-\frac{\varphi_o^{lim} - \varphi_o}{2V_{th}}\right), \tag{58}$$

and

$$\varphi_o^{lim} = V_{th} \ln\left(\frac{2\pi^2 \varepsilon_{si} N_a V_{th}}{qn_i^2 a^2}\right) \tag{59}$$

is the limiting value of φ_o for large V_{gs} (applies also to the SqG MOSFET). We find the following implicit expression for the center potential by considering the potential distribution from gate-to-gate, including the linear potential drops across the gate insulator,

$$V_{gs} - V_{FB} - \varphi_o = \frac{2V_{th}}{\tau}\left\{2\kappa\left(\frac{ha}{2}\right)\tan\left(\frac{ha}{2}\right) - \ln\left[\cos\left(\frac{ha}{2}\right)\right]\right\}, \qquad (60)$$

where $\tau = 1$ for the DG device and $\kappa = \varepsilon_{si}t_{ox}/\varepsilon_{ox}a$ is a structural parameter. As φ_o approaches the limiting value φ_o^{lim}, the asymptotic behavior of (60) can be expressed analytically as follows for $\varphi_o^{lim} - \varphi_o < 5\,\mathrm{mV}$,

$$\varphi_o = \varphi_o^{lim} - \frac{8\kappa V_{th}^2/\tau}{V_{gs} - V_o}, \qquad (61)$$

where

$$V_o = V_{FB} + \varphi_o^{lim} + \left(2V_{th}/\tau\right)\left[1 - 2\kappa - \ln\left(\pi/2\right)\right]. \qquad (62)$$

As observed above, the last term on the right-hand side of (61) is much smaller than the other terms in strong inversion. The same goes for the last term on the right-hand side of (62).

4.2.2. Strong inversion electrostatics in SqG MOSFETs

In SqG MOSFETs, Poisson's equation for the center potential becomes

$$\left[\frac{\partial^2\varphi(x,y,z)}{\partial x^2} + \frac{\partial^2\varphi(x,y,z)}{\partial y^2}\right]_{x=y=0} \equiv 2\frac{\partial^2\varphi(x,y,z)}{\partial x^2}\bigg|_{x=y=0} = -\frac{q}{\varepsilon_{si}}\frac{n_i^2}{N_a}\exp\left(\frac{\varphi_o}{V_{th}}\right). \qquad (63)$$

We note that this expression has the same form as (56) for the DG device, except for the factor of 2 originating from the equivalence of the x and y axes. We emphasize, however, that (63) strictly applies only at the center of the (x,y) cross section. In addition, the potential is also fixed at the gate boundary. Therefore, we expect that the potential along the x or y axis of the SqG MOSFET can be derived with a fairly good approximation from that of the DG device, i.e.,

$$\varphi(x) = \varphi_o - \frac{2V_{th}}{\tau}\ln\left[\cos(hx)\right], \qquad (64)$$

and likewise in the y direction. Here we should ideally expect that $\tau = 2$ because of the fourfold symmetry. However, since (64) with $\tau = 2$ is not quite precise everywhere in the SqG silicon body, we have used the parameter τ to fine-tune the overall precision along the x and y directions. But, as indicated in conjunction with (61) and (62), the terms containing τ are quite small, making (64) relatively insensitive to this parameter. However, the value $\tau = 1.5$ seems to be the most appropriate.

Since the potential distributions are identical for the x and y directions, we utilize the symmetry properties of the device to construct a complete isomorphic solution for the silicon body of the form

$$\varphi(x, y) = \varphi_o - \frac{2V_{th}}{\tau}\left\{\ln\left[\cos(hx)\cos(hy)\right] - \eta\frac{\ln\left[\cos(hx)\right]\ln\left[\cos(hy)\right]}{\ln\left[\cos(ha/2)\right]}\right\}, \qquad (65)$$

where the parameter $\eta \approx 0.638$ serves to optimize the overall solution in the (x,y) cross-section, in particular along the diagonal directions.

Fig. 50 shows that the model based on the above expressions compares very well numerical simulations along high-symmetry directions in the central (x,y) plane for $V_{gs} = 0.7$ V and 0.8 V. We notice the increased potential in the body corners, which gives an enhanced electron concentration in these regions. In Fig. 51 is shown the modeled potential distribution along the silicon-insulator interface, and Fig. 52 shows a 3D model rendering of the full potential distribution in the (x,y) cross section for $V_{gs} = 0.8$ V.

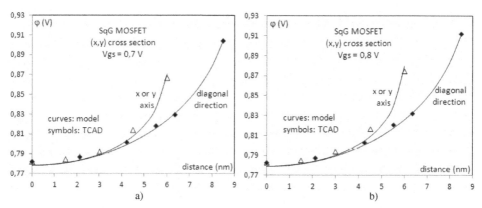

Fig. 49. Comparisons between modeled and numerically simulated strong inversion potential distributions in the (x,y) cross section along x direction and along the diagonal for (a) $V_{gs} = 0.7$ V and (b) $V_{gs} = 0.8$ V.

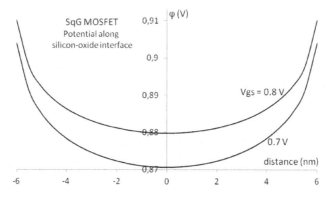

Fig. 50. Modeled potential distribution of SqG MOSFET along the silicon/oxide interface for $V_{gs} = 0.7$ V and 0.8 V.

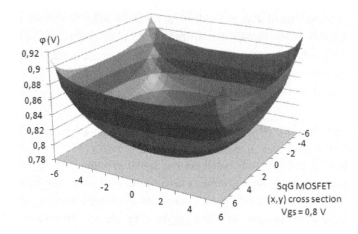

Fig. 51. Modeled potential distributions of SqG MOSFET in the (x,y) cross section inside the silicon body for $V_{gs} = 0.8$ V. (After Ref. 34.)

4.2.3. *Strong inversion charge, drain current and capacitances*

The charge carrier density per unit length in the z direction of the SqG device is obtained by integrating the perpendicular displacement at the silicon-oxide interface over this interface. Using (59), this result can be expressed as

$$Q_s = -8\varepsilon_{si}\int_0^{a/2} \frac{d\varphi(x,y)}{dx}\bigg|_{x=a/2} dy = Q_o\left[1-F_o\right], \tag{66}$$

where

$$Q_o = -8\varepsilon_{si}\frac{2V_{th}}{\tau}\left(\frac{ha}{2}\right)\tan\left(\frac{ha}{2}\right), \tag{67}$$

$$F_o = \frac{2\eta}{ha}\frac{\int_0^{ha/2}\ln\left[\cos(s)\right]ds}{\ln\left[\cos(ha/2)\right]} = \eta\left\{1+\frac{2}{ha}\frac{\int_0^{ha/2}s\tan(s)ds}{\ln\left[\cos(ha/2)\right]}\right\}. \tag{68}$$

Considering the strong inversion asymptotic region, where $ha/2$ is quite close to $\pi/2$, we can write $ha/2 \approx \pi/2$, $\cos(ha/2) \approx (\pi/2)(1 - ha/\pi)$, $\ln[\cos(ha/2)] \approx \ln(1 - ha/\pi))$, and $\tan(ha/2) \approx (2/\pi)/(1 - ha/\pi) \approx 4V_{th}/(\pi\Delta\varphi_o)$, where $\Delta\varphi_o = \varphi_o^{lim} - \varphi_o = (8\kappa V_{th}^2/\tau)/(V_{gs}-\varphi_o^{lim}-V_o)$ according to (61). Applying this to (67) and (68), we find that $F_o \to 0$ and

$$Q_s \approx Q_o \approx -\frac{4\varepsilon_{si}}{\kappa}\left(V_{gs}-V_o\right), \tag{69}$$

So far we have only considered the electrostatics a zero drain bias. To include drain bias, we simply add the quasi-Fermi potential $V_F(z)$ to φ_o and, by inference, also to φ_o^{lim} in (65). Here, $V_F \approx 0$ at the source side and $V = V_{ds}$ at the drain side of the effective channel.

By introducing the quasi-Fermi potential this way, the parameter h becomes independent of $V_F(z)$. Hence, (65) and (69) can be rewritten as

$$\varphi(x,y,z) = \varphi_o + V_F(z) - \frac{2V_{th}}{\tau}\left\{\ln\left[\cos(hx)\cos(hy)\right] - \eta\frac{\ln\left[\cos(hx)\right]\ln\left[\cos(hy)\right]}{\ln\left[\cos(ha/2)\right]}\right\}, \quad (70)$$

and

$$Q_s \approx -\frac{4\varepsilon_{si}}{\kappa}\left[\left(V_{gs} - V_o\right) - V_F\right], \quad (71)$$

respectively (note that V_o includes φ_o^{lim} as an additive term, see (62)).

The strong inversion drain current of the SqG MOSFET is determined self-consistently from the above expressions by writing

$$I_d = -Q_s(z)v = \frac{4\varepsilon_{si}\mu_o}{\kappa}\left[\left(V_{gs} - V_o\right) - V_F\right]\frac{dV_F}{dz}, \quad (72)$$

where the carrier velocity v is based on the constant mobility approximation. Using the boundary conditions for $V_F(z)$, (72) can be solved to give the following result for the triode region,[34]

$$I_d = \frac{4\varepsilon_{si}\mu_o}{\kappa L_{eff}}\left[\left(V_{gs} - V_o\right)V_{ds} - \frac{1}{2}V_{ds}^2\right], \quad (73)$$

where the effective channel length L_{eff} will be reduced compared to the geometrical gate length L by the channel length modulation mechanism (CLM), similar to that of the classical MOSFETs. This gives the saturation voltage $V_{sat} = V_{gs} - V_o$ and the saturation current becomes

$$I_{dsat} = \frac{2\varepsilon_{si}\mu_o}{\kappa L_{eff}}\left(V_{gs} - V_o\right)^2. \quad (74)$$

The strong inversion threshold voltage is $V_{Ts} = V_o$. We emphasize that the present saturation and threshold voltages are applicable only to the strong inversion regime.

Saturation of the carrier velocity can also be included in I_d by using the expression $\mu_n = \mu_o/(1 + E/E_s)$, where $E = -dV_F/dz$ and E_s is the saturation field. In this case we find,

$$I_d = \frac{4\varepsilon_{si}\mu_o|E_s|}{\kappa}\frac{\left(V_{gs} - V_o\right)V_{ds} - V_{ds}^2/2}{V_{ds} + L_{eff}|E_s|}, \quad (75)$$

To derive the intrinsic capacitances in the linear region, we first have to integrate Q_s along the channel. Applying (71) and (74), we find

$$Q_g = -Q_{ch} = -\int_0^{L_{eff}} Q_s dz = \left(\frac{4\varepsilon_{si}}{\kappa}\right)^2 \frac{\mu_o}{I_d} \int_0^{V_{ds}} \left[(V_{gs} - V_o) - V_F\right]^2 dV_F$$

$$= \frac{4\varepsilon_{si} L_{eff}}{\kappa} \frac{(V_{gs} - V_o)^2 - (V_{gs} - V_o)V_{ds} + \frac{1}{3}V_{ds}^2}{(V_{gs} - V_o) - \frac{1}{2}V_{ds}} \tag{76}$$

From this we can obtain analytical expressions for the primary intrinsic capacitances,

$$C_{gg} = \frac{\partial Q_g}{\partial V_s}\bigg|_{V_s, V_d} = C_o \left\{ 1 - \frac{1}{3}\left[\frac{V_{ds}}{2(V_{gs} - V_o) - V_{ds}}\right]^2 \right\}, \tag{77}$$

$$C_{gs} = \frac{\partial Q_g}{\partial V_s}\bigg|_{V_g, V_d} = \frac{2}{3}C_o \left\{ 1 - \left[\frac{(V_{gs} - V_o) - V_{ds}}{2(V_{gs} - V_o) - V_{ds}}\right]^2 \right\}, \tag{78}$$

$$C_{gd} = \frac{\partial Q_g}{\partial V_s}\bigg|_{V_g, V_s} = \frac{2}{3}C_o \left\{ 1 - \left[\frac{V_{gs} - V_o}{2(V_{gs} - V_o) - V_{ds}}\right]^2 \right\}. \tag{79}$$

where $C_o = 4\varepsilon_{si} L_{eff}/\kappa$ is the gate capacitance at zero drain bias. We note that C_{gs} and C_{gd} are the well-known Meyer capacitances for long-channel MOSFETs. [48] To obtain a complete charge conserving capacitance model, we also have to calculate the charges Q_s and Q_d assigned to source and drain, respectively, using a suitable charge partitioning scheme. This results in a total of nine self- and trans-capacitances, of which four are independent. However, if the channel is sufficiently long, we can assume that C_{ds} and C_{sd} are negligible, in which case C_{ss}, C_{sg}, C_{dd}, and C_{dg} can be determined from those of (77)-(79).

4.2.4. *Strong inversion electrostatics in other multigate devices MOSFETs*

The above analysis of strong inversion operation can readily be extended to include other multigate MOSFETs, such as RecG, trigate and FinFet devices, similar to what was discussed in subthreshold operation, see Section 4.1.3. Again, the devices of rectangular cross section with an aspect ratio of more than a certain value will develop a DG-like section about the central (x,z) plane. However, since the screening length is now quite small owing to the inversion charge, the ReG device can be considered a DG device capped with one half of a SqG device at each end, as indicated for the subthreshold case in the inset of Fig. 46. The trigate device is basically the symmetric half of a RecG device (see inset in Fig. 47). The FinFET can be either of these devices, typically with an extended aspect ratio (tall and thin device) and behaving primarily as a DG device, with some influence from the top and bottom sections, depending on how they are terminated.

In conclusion, the modeling techniques presented above therefore forms an excellent basis for a unified modeling framework, covering both subthreshold and strong inversion operation, for a wide range of multigate devices of high technological interest.

5. Modeling of Quantum Mechanical Effects in MugFETs

Modeling of the quantum effects in the nanoscale semiconductor devices with width and/or length dimensions less than about 10 nm is imperative for a precise calculation of the charge density, charge potential and prediction of the transport characteristics. Unlike single-gate devices, MugFETs suffer from the strong structural confinement along with the electronic confinement at higher gate electric fields. At low gate fields (subthreshold and near threshold), structural confinement is more dominant and at high gate fields (strong inversion) electronic confinement takes over. Quantization models based on triangular wells – used in single-gate devices – can be applied in the strong inversion regime, but new modeling methodology is required for the subthreshold and near-threshold regime, where the energy well is centered in the body interior. In the following sections Here, we present a few modeling examples of the quantum effects in DG and GAA devices.

5.1.1. *Quantum confinement modeling in DG MOSFETs*

DG MOSFETs suffer mainly 1D confinement along the G-G direction. Wave penetration into the gate insulator is neglected. The corresponding Schrödinger is then given as

$$-\frac{\hbar^2}{2m_y^*}\frac{d^2\psi(y)}{dy^2}+q\varphi\,(y)\psi(y)=E_T\psi(y),\tag{80}$$

where $\psi(y)$ is the electron eigenfunction, E_T is the total eigenvalue including the conduction band edge relative to the intrinsic Fermi level and the subband offset, $\varphi(y)$ is the total potential along G-G axis, and m_y^* is the effective mass along the y-direction. Here $y = 0$ corresponds to D-S symmetry line (x axis). To solve the Schrödinger equation, we approximate the total potential in the subthreshold and near-threshold regimes by a parabolic function

$$\varphi(x,y)\approx\hat{\varphi}(x)\left[1-\left(\frac{2y}{a'}\right)^2\right]+V_{gs}-V_{FB},\tag{81}$$

where $\hat{\varphi}(x)$ is the potential at position x on the S-D symmetry axis referred to that of the gate-silicon interface of the extended body and $a' = a + 2t'_{ox}$ (see Section 2.1). This approximation is justified in subthreshold when we have volume inversion, but fails in strong inversion where the charge is mainly confined to triangular wells at the silicon insulator interface.

Using (81), the corresponding Schrödinger equation along the y-axis becomes

$$-\frac{\hbar^2}{2m_y^*}\frac{d^2\psi(y)}{dy^2}+q\varphi(x)\left(\frac{2y}{a}\right)^2\psi(y)=E\psi(y),\qquad(82)$$

where E is now the subband offset from the conduction band edge. This equation generates two different sets of eigenstates, one for each effective mass. The lower eigenvalues correspond to the higher effective mass ($m_l^* = 0.98m_o$, known as unprimed bands in the literature) and the higher eigenvalues for the lower effective mass ($m_t^* = 0.19m_o$, known as primed bands). Equation (82) can be transformed into the standard equation of the parabolic cylinder functions[36,49].

$$\frac{d^2\psi(t)}{dt^2}-\left(\frac{t^2}{4}+v\right)\psi(t)=0,\qquad(83)$$

where $t=y\left[32m_y^*q\hat{\varphi}(x)/(a\hbar)^2\right]^{1/4}$ and $v=E/\sqrt{8q\hat{\varphi}(x)\hbar^2/m_y^*a^2}$. This equation has both even and odd solutions. As the potential is symmetric about $y = 0$, the wave function must have a definite parity and thus $\psi(t)$ is given by two different polynomials with even (e) and odd (o) terms

$$\psi_{e,o}(t)=\begin{cases}A_e\left(1+\sum_{n(even)=2}^{\infty}\frac{c_n t^n}{n!}\right),\\[2ex]A_o\left(1+\sum_{n(odd)=3}^{\infty}\frac{c_n t^n}{n!}\right)\end{cases}\qquad(84)$$

where A_e and A_o are normalization constants and the coefficients c_n are determined from a recursive relation[49]. Depending upon the value of t, these eigenvalues can be determined by truncating the above polynomials at some higher term and then finding the numerical solution. For ultra-thin devices operating in subthreshold and near-threshold regimes, the first four terms of the polynomials give sufficient accuracy in the calculation of the lowest subbands. Since higher subbands are less affected by the potential perturbation, they can be better approximated by the eigenvalues of a particle in a box. For devices operating in the near-threshold regime, the eigenvalues and eigenfunctions approach structure dependent values. In this case, the eigenfunction and eigenvalue corresponding to the first unprimed subband are given as[50]

$$\psi_1(y)=\sqrt{\frac{15}{8a}}\left[1-\left(\frac{2y}{a}\right)^2\right],\qquad E_1=\frac{4\hbar^2}{m_l^*a^2}\qquad(85)$$

Figure 52 shows the variation of the electron density along the y-axis for a device with $a = 3$ nm for $V_{ds} = V_{gs} = 0$ V. The material and other geometric parameters used are specified in Section 1 and Section 2.2, respectively, unless indicated otherwise.

Near and above threshold, the effect of the charge carriers on the device electrostatics become important and thus the total body potential is given as

$$\varphi(x,y)=\varphi^{LP}(x,y)+\varphi^Q(x,y),\qquad(86)$$

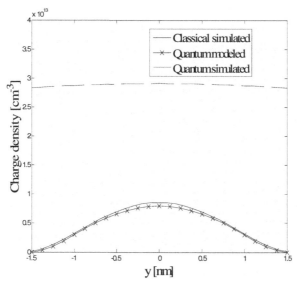

Fig. 52. Comparison of modeled and numerically simulated electron density along the *y* axis for a device with *a* = 3 nm at $V_{gs} = V_{ds} = 0$ V. Both quantum mechanical and classical simulation results are shown.

where $\varphi^Q(x,y)$ is the potential due to the charge carriers. Ideally, a complete solution of the 2D Poisson's equation must be obtained for the entire device, which is prohibitively complex. To circumvent this issue, we basically solve the 1D Poisson's equation in the central region near the top of the barrier. For all practical devices, this is a reasonable assumption. We here demonstrate the solution for the top of the barrier, which is at the device center for $V_{ds} = 0$ V.

Assuming that the curvature of $\varphi^c(x,y)$ in the *x* direction is negligible at the device center $(x = x_o)$, the 1D quantum Poisson's equation along the gate-to-gate axis for $V_{ds} = 0$ is given as

$$\frac{d^2\varphi^Q(x_o,y)}{dy^2} = \frac{q}{\varepsilon_{si}} \sum_{valleys} g_v \sum_j \ln\left[1+\exp\left(\frac{E_F(x_o)-E_j(x_o)}{k_BT}\right)\right] \left|\psi_{j(y)}\right|^2, \tag{87}$$

where N_{2D} is the 2D effective density of states, g_v denotes the valley degeneracy, and $E_F(x_o)$ is the Fermi level at the device center, which is the same as that of the source and drain since $V_{ds} = 0$ V. This equation is still quite complex and it is difficult to obtain an analytical solution. Therefore, we make additional assumptions: Firstly, we assume that only the first subband is occupied. This is quite justified in ultra-thin body (UTB) devices since the energy difference between different subbands increase as the silicon thickness is reduced. For relatively higher thickness, the single subband charge density is replaced by the total charge density after the solution is obtained. Secondly, we assume a non-degenerate carrier statistics near the threshold voltage. At the onset of threshold, the the charge density is still low and thus the non-degenerate statistics gives a negligible error. Using (87) along with the above assumptions, the resulting Poisson equation along the y axis becomes[49]

$$\frac{d^2\varphi^Q(x_o,y)}{dy^2} = \frac{15qN_{2D}}{8a\varepsilon_{si}}\exp\left(\frac{E_F(x_o)-E_j(x_o)}{k_BT}\right)\left[1-\left(\frac{2y}{a}\right)^2\right]^2, \quad (88)$$

The resulting potential contribution from the inversion charge at the device center then becomes

$$\varphi_o^Q = \frac{qN_{2D}t_{si}}{8\varepsilon_{si}}\left(\frac{11}{8}+\frac{4t_{ox}}{t_{si}}\right)\exp\left[\frac{E_g/2+E_j(x_0)+q(\varphi_b-\varphi_i-\varphi_o^{LP}(x_o,0)-\varphi_o^Q(x_o,0))}{k_BT}\right], \quad (89)$$

where E_g is the energy band gap of silicon, φ_b is the difference between the Fermi potentials of doped and undoped silicon, and φ_i is the difference between the intrinsic and the middle potential of the band gap. We observe that (89) can be expressed in the form of a Lambert function. Once the center potential is derived, a complete solution can be obtained using the parabolic form in (81).

Figure 53 shows the total modeled potential at the device center, compared with numerical simulations for $0.2V \le V_{gs} \le 0.36$ V, which corresponds to the near-threshold regime. The device thickness here is 6 nm. A similar procedure can be used to correct the threshold voltage in presence of quantum confinement[51].

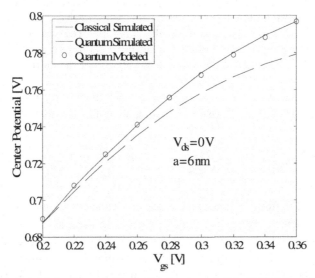

Fig. 53. Variation of modeled and numerically simulated center potential with the gate voltage for a device with $a = 6$ nm DG MOSFET at $V_{ds} = 0$ V. Both quantum mechanical and classical simulation results are shown. (After Ref. 49.)

From the quantum potential modeling above, it is possible to calculate the subthreshold current in ultra thin DG MOSFETs, following the procedure indicated in Section 2.2.4. Figure 54 shows a comparison of modeled and numerically simulated subthreshold I_{dsub}-V_{gs} characteristics for two devices with $a = 3$ nm and 6 nm.

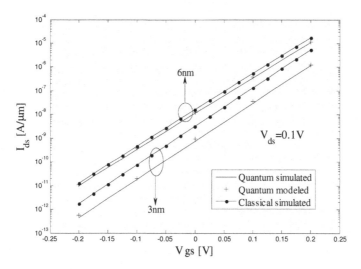

Fig. 54. Comparison of modeled and numerically simulated subthreshold I_{dsub}-V_{gs} characteristics for DG MOSFETs with a = 3 nm and 6 nm. Both quantum mechanical and classical simulation results are shown. (After Ref. 49.)

5.2. *Quantum confinement modeling in CirG MOSFETs*

Compared to the DG MOSFET, the cylindrical GAA device suffers from confinement in the two dimensions, the (x,y) plane perpendicular to the cylinder axis (z axis). Again using a parabolic approximation for the potential distribution in the radial direction, the Schrödinger equation for this case becomes

$$\left\{-\frac{\hbar^2}{2m_y^*}\left[\frac{1}{\rho}\frac{\partial}{\partial\rho}\left(\rho\frac{\partial}{\partial\rho}\right)+\frac{1}{\rho^2}\frac{\partial^2}{\partial\phi^2}\right]+q\hat{\phi}(x)\left(\frac{2\rho}{a}\right)^2\right\}\psi(\rho,\phi)=E\psi(\rho,\phi),\qquad(90)$$

where ϕ is the angle in the azimuthal direction. Using separation of the variables, i.e. assuming that the total wavefunction is a product of radial and azimuthal wavefunctions, the above equation can be separated into two different equations

$$\frac{\rho^2}{R(\rho)}\frac{d^2R(\rho)}{d\rho^2}+\frac{\rho}{R(\rho)}\frac{dR(\rho)}{d\rho}+\alpha\rho^2-\lambda\rho^4=m^2,\qquad(91)$$

$$\frac{1}{\Phi(\phi)}\frac{d^2\Phi(\phi)}{d\phi^2}=-m^2,\qquad(92)$$

where m is the magnetic quantum number, $\alpha=\sqrt{2m*E/\hbar^2}$, and $\lambda=\sqrt{8m*q\phi^Q/(\hbar a')^2}$. Here, $a' = t_{si} + 2t'_{ox}$ is the diameter of the extended body, as before. Hence, the complete solution of (88) is the product of the solutions in (91) and (92), and is given in terms of associated Laguerre functions

$$\psi(\rho,\phi) = Ce^{-\lambda\rho^2/2} 2^{(|m|+1)/2} \rho^{|m|} L_\beta^{|m|}\left(\lambda\rho^2\right)e^{jm\phi}, \tag{93}$$

where C is a normalization constant and the index β is given as $\beta = \left[\alpha^2 - 2(|m|+1)\lambda\right]/(4\lambda)$.

Figure 55 shows the modeled probability density $|\psi(\rho)|^2$, compared with numerical simulations[52]. The CirG device considered has a gate length $L = 15$ nm, and a silicon diameter $a = 12$ nm. All other parameters are as for the DG device in Section 5.1.1.

Since the wavefunction should vanish at the boundaries, this imposes the condition

$$L_\beta^{|m|}\left(\lambda a^2/4\right) = 0. \tag{94}$$

Thus the eigenvalues are given as

$$E_{m,n} = \hbar\omega(2\beta + |m| + 1), \tag{95}$$

where the subscript n denotes the nth root of (92) for the mth magnetic quantum number, β is the real number for which (94) is satisfied, and $\omega = \sqrt{8q\varphi^\varrho/(m*a^2)}$. In (95), m is an integer and β is generally a non-integer. All the energy levels with non-zero m are doubly degenerate, which corresponds to a clockwise and an anti-clockwise polarization of the eigenfunctions.

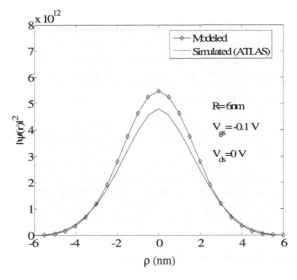

Fig. 55. Modeled electron probability density at the device center, compared with the numerical simulations t_{si} = 3 nm and 6 nm. (After Ref. 51.)

Acknowledgment

The work was carried out with support by the European Commission under Grant Agreement 218255 (COMON), and the Norwegian Research Council under contract 970141669 (MUSIC).

References

[1] J. P. Colinge, "Multi-gate SOI MOSFETs," *Microelectronics Eng.*, vol. 84, pp. 2071-2076, 2007.

[2] Y. Taur, "An Analytical Solution to a Double-Gate MOSFET with Undoped Body," *IEEE Electron Device Lett.*, vol. 21, no. 5, pp. 245–247, 2000.

[3] Y. Taur, X. Liang, W. Wang, and H. Lu, "A Continuous, Analytic Drain-Current Model for DG-MOSFETs," *IEEE Electron Device Lett.*, vol. 25, no. 2, pp. 107-109, 2004.

[4] A. Ortiz-Conde , F. J. Garcia Sanchez, and J. Muci. "Rigorous Analytic Solution for the Drain Current of Undoped Symmetric Dual-Gate MOSFETs," *Solid-State Electron.*, vol.49, pp. 640-647, 2005.

[5] D. Jiménez, B. Iñiguez, J. Suñé, L. F. Marsal, J. Pallarès, J. Roig, and D. Flores, "Continuous Analytic Current-Voltage Model for Surrounding-Gate MOSFETs," *IEEE Electron Device Lett.*, vol. 25, no. 8, pp. 571-573, 2004.

[6] B. Iñiguez, D. Jiménez, J. Roig, H. A. Hamid, L. F. Marsal, and J. Pallarès, "Explicit Continuous Model for Long-Channel Undoped Surrounding Gate MOSFETs," *IEEE Trans. Electron Devices*, vol. 52, pp. 1868-1873, August 2005.

[7] A. Ortiz-Conde , F. J. Garcia Sanchez, J. Muci, S. Malobabic, and J. J. Liou, "A Review of Core Compact Models for Double-Gate SOI MOSFETs," *IEEE Trans. Electron Devices*, vol. 54, pp. 131-139, 2007.

[8] W. Z. Shangguan, X. Zhou, K. Chandrasekaran, Z. Zhu, S. C. Rustagi, S.B. Chiah, and G. H. See, "Surface Potential Solution for Generic Undoped MOSFETs with Two Gates," *IEEE Trans. Electron Devices*, vol. 54, pp. 169-171, 2007.

[9] F. Liu, J. He, L. Zhang, J. Zhang, J. Hu, C. Ma, and M. Chan, "A Charge Based Model for Long-Channel Cylindrical Surrounding Gate MOSFET from Intrinsic to Heavily Doped Channel," *IEEE Trans. Electron Devices*, vol. 55, no. 8, pp. 2187–2194, Aug. 2008.

[10] B. Yu, J. He, J. Song, Y. Yuan, and Y. Taur, "A Unified Analytic Drain-Current model for Multigate MOSFETs," *IEEE Trans. Electron Devices*, vol. 55, no. 8, pp. 2157–2163, Aug. 2008.

[11] S. K. Vishvakarma, A. K. Saxena, and S. Dasgupta, "Two Dimensional Analytical Potential Modeling of Nanoscale Fully Depleted Metal Gate Double-Gate MOSFET," *J. Nanoelectronics Optoelectronics*, vol. 3, no. 3,pp. 297–306, Dec. 2008.

[12] Y. P. Liang and Y. Taur, "A 2-D Analytical Solution for SCEs in DG MOSFETs," *IEEE Trans. Electron Devices*, vol. 51, no. 9, pp. 1385-1391, 2004.

[13] D. J. Frank, Y. Taur, H. –S. P. Wong, "Generalized Scale Length for Two-Dimensional Effects in MOSFETs," *IEEE Electron Device Lett.*, vol. 19, pp. 385-387, 1998.

[14] S. –H. Oh, D. Monroe and J. M. Hergenrother, "Analytic Description of Short-Channel Effects in Fully-Depleted Double-Gate and Cylindrical, Surrounding-Gate MOSFETs," *IEEE Electron Device Lett.*, vol. 21, no. 9, pp. 445–447, 2000.

[15] E. Weber, *Electromagnetic Fields,* vol. 1, New York: Wiley, 1950.

[16] A. Klös and A. Kostka, "A New Analytical Method of Solving 2D Poisson's Equation in MOS Devices Applied to Threshold Voltage and Subthreshold Modeling," *Solid-State Electron.*, vol. 39, pp. 1761-1775, 1996.

[17] J. Østhaug, T. A. Fjeldly, and B. Iñíguez, "Closed-Form 2D Modeling of Sub-100 nm MOSFETs in the Subthreshold Regime," *J. Telecommun. and Inf. Technol.*, no. 1, pp. 70-79, 2004.

[18] S. Kolberg and T. A. Fjeldly, "2-D Modeling of Nanoscale DG SOI MOSFETs," *J. Comput. Electron.*, vol. 5, no. 2/3, pp. 217-222, 2006.

[19] S. Kolberg and T. A. Fjeldly, "2D Modeling of Nanoscale Double Gate Silicon-on-Insulator MOSFETs Using Conformal Mapping," *Phys. Scripta*, vol. T126, pp. 54-60, 2006.

20   S. Kolberg, T. A. Fjeldly, and B. Iñiguez, "Self-Consistent 2D Compact Model for Nanoscale Double Gate MOSFETs," *Lecture Notes in Comp. Science,* Springer-Verlag, Berlin, vol. 3994, pp. 607-614, 2006.

21   B. Iñiguez, T. A. Fjeldly, A. Lazaro, F. Danneville, and M. J. Deen, "Compact Modeling Solutions for Double Gate and Gate-All-Around MOSFETs," *IEEE Trans. Electron Devices,* vol. 53, no. 9, pp. 2128-2142, 2006.

22   H. Børli, S. Kolberg, and T. A. Fjeldly, "Compact Modeling Framework for Short-Channel DG and GAA MOSFETs," *Proc. NSTI-Nanotech,* San Francisco, CA, May 20-14, 2007, vol. 3, pp. 505-509.

23   S. Kolberg, *Modeling of Electrostatics and Drain Current in Nanoscale Double-Gate MOSFETs,* PhD thesis, Norwegian University of Science and Technology, 2007.

24   H. Børli, S. Kolberg, T. A. Fjeldly, and B. Iniguez, "Precise Modeling Framework for Short-Channel Double-Gate and Gate-All-Around MOSFETs," *IEEE Trans. Electron Devices,* vol. 55, no. 10, pp. 2678–2686, Oct. 2008.

25   H. Børli, S. Kolberg, and T. A. Fjeldly, "Capacitance Modeling of Short-Channel Double-Gate MOSFETs," *Solid State Electronics,* Vol. 52, pp. 1486-1490, 2008.

26   H. Børli, S. Kolberg, and T. A. Fjeldly, "Physics Based Current and Capacitance Model of Short-Channel Double Gate and Gate-All-Around MOSFETs," *Proc. IEEE Int. Conf. on Nanoelectr. (INEC),* March 24-27, 2008, Shanghai, China, pp. 844-849.

27   H. Børli, *Modeling of Drain Current and Intrinsic Capacitances in Nanoscale Double-Gate and Gate-All-Around MOSFETs,* PhD thesis, Norwegian University of Science and Technology, 2008.

28   A. Kloes, M. Weidemann, D. Goebel, and B. T. Bosworth, "Three-Dimensional Closed-Form Model for Potential Barrier in Undoped FinFETs Resulting in Equations for V_T and Subthreshold Slope," *IEEE Trans. Electron Devices,* vol. 55, no. 12, pp. 3467–2475, Dec. 2008

29   H. A. El Hamid, B. Iñiguez, and J. Roig, "Analytical Model of the Threshold Voltage and Subthreshold Swing of Undoped Cylindrical Gate-Allaround-Based MOSFETs," *IEEE Trans. Electron Devices,* vol. 54, no. 3, pp. 572–579, Mar. 2007.

30   S. Kolberg, H. Børli, and T. A. Fjeldly, "Modeling, Verification and Comparison of Short-Channel Double Gate and Gate-All-Around MOSFETs," *J. Math. Comput. Simul.,* vol. 79, pp. 1107-1115, 2008.

31   T. A. Fjeldly and H. Børli, "2-D Modeling of Nanoscale Multigate MOSFETs," *Proc. Int. Conf. on Solid-State and Integr. Circ. Techn. (ICSICT),* Beijing, China, Oct. 20-23, 2008, paper B5.1.

32   S. K. Vishvakarma, U. Monga, and T. A. Fjeldly, "Analytical Modeling of the Subthreshold Electrostatics of Nanoscale GAA Square Gate MOSFETs," *Proc. NSTI-Nanotech,* Anaheim, CA, June 21-24, 2010, vol. 2, pp. 789-792.

33   S. K. Vishvakarma, U. Monga, and T. A. Fjeldly, "Unified Analytical Modeling of Nanoscale GAA MOSFETs," *Proc. Int. Conf. on Solid-State and Integr. Circ. Techn. (ICSICT),* Shanghai, China, Nov. 1-4, 2010, vol. 3, pp. 1733-1736.

34   T. A. Fjeldly and U. Monga, "Unified Modeling of Multigate MOSFETs Based on Isomorphic Modeling Principles", *NIST, Nanotech,* Santa Clara, MA, June 18-21, 2012, vol. 2, pp. 756-761.

35   Y. Omura and K. Izumi, "Quantum Mechanical Influences on Short-Channel Effects in Ultra-Thin MOSFET/SIMOX Devices," *IEEE Electron Device Lett.,* vol. 17, no. 6, pp. 300-3003, 1996.

36   H-S. P. Wong, D. J. Frank, and P. M. Solomon, "Device Design Considerations for Double-Gate, Ground-Plane, and Single-Gated Ultra-Thin SOI MOSFET's at the 25 nm Channel Length Generation," *IEDM Tech. Dig.,* pp. 407-410, Dec. 1998.

37   M. Abramowitz and I. A. Stegun, *Handbook of Mathematical Functions.* New York: Dover Publications, 1964.

[38] U. Monga and T. A. Fjeldly, "Compact Subthreshold Current Modeling of Short-Channel Nanoscale Double-Gate MOSFET," *IEEE Trans. Electron Devices*, vol. 56, pp. 1533-1537, no. 7, July 2009.

[39] H. Børli, K. Vinkenes, and T. A. Fjeldly, "Physics Based Capacitance Modeling of Short-Channel Double-Gate MOSFETs," *Phys. Stat. Sol. (c)*, vol. 5, no. 12, pp. 3643-3646, 2008.

[40] H. C. Pao and C. T. Sah, "Effects of Diffusion Current on Characteristics of Metal-Oxide (Insulator) Semiconductor Transistors," *Solid State Electron.*, vol. 9, no. 10, pp. 927–937, Oct. 1966.

[41] U. Monga, H. Børli, and T. A. Fjeldly, "Compact Subthreshold Current and Capacitance Modeling of Short-Channel Double-Gate MOSFET," *Math.Comp. Mod.*, vol. 51, no. 7-8, pp. 901-907, 2010.

[42] U. Monga and T. A. Fjeldly, "Compact Subthreshold Slope Modeling of Short-Channel Double-Gate MOSFET," *Electronics Lett.*, vol. 45, no. 9, pp. 476-478, 2009.

[43] S. Kolberg. H. Børli, and T. A. Fjeldly, "Compact Current Modeling of Short-Channel Multiple Gate MOSFETs," *Phys. Stat. Sol. (c)*, vol. 5, no. 12, pp. pp. 3609-3612, 2008.

[44] J. P. Colinge, "The SOI MOSFET: from Single Gate to Multigate," in *FinFETs and Other Multi-Gate Transistors*, J. P Colinge, Editor. Berlin: Springer-Verlag, 2008.

[45] K. Suzuki, Y. Tosaka, T. Tanaka, H. Horie, and Y. Arimoto, "Scaling Theory for Double-Gate SOI MOSFETs," *IEEE Trans. Electron Devices*, vol. 40, pp. 2326-2329, 1993.

[46] C. P. Auth and J. D. Plummer, "Scaling Theory for Cylindrical Fully Depleted, Surrounding Gate MOSFET, "*IEEE Electron Device Lett.*, vol. 18, pp. 74-76, 1997.

[47] T.A. Fjeldly and U. Monga, "Compact Isomorphic Modeling of Rectangular Gate and Trigate MOSFETs", *Proc. NSTI-Nanotech*, Boston, MA, June 13-16, 2011, vol. 2, pp. 732-737, 2011.

[48] T. A. Fjeldly, T. Ytterdal, and M. Shur, *Introduction to Device Modeling and Circuit Simulation*, New York: Wiley, 1998.

[49] A. Consortini and B. R. Frieden, *Italian Physical Society*, vol. 35, No. 2, 1976.

[50] U. Monga, and T. A. Fjeldly, "Compact Quantum Modeling Framework for Nanoscale Double-Gate MOSFETs," *Proc. NSTI-Nanotech*, Houston, TX, May 3-7, 2009, vol. 3, pp. 580-583.

[51] U. Monga, and T. A. Fjeldly, "Subthreshold Quantum Ballistic Current and Quantum Threshold Voltage Modeling for Nanoscale Nanoscale FinFETs," *Proc. NSTI-Nanotech*, Anaheim, CA, June 21-24, 2010, vol. 2, pp. 769-772.

[52] U. Monga, and T. A. Fjeldly, "Modelling of Quantum Ballistic Cylindrical Nanowire MOSFETs in the Subthreshold Regime," *Phys. Scr.*, vol. T141, paper 014016, 2010.

COMPACT MODELING OF DOUBLE AND TRI-GATE MOSFETs

BENJAMIN IÑIGUEZ

DEEEA/ETSE, Universitat Rovira I Virgili, 26 Avinguda dels Països Catalans,
Tarragona, 43007, Spain
benjamin.iniguez@urv.cat

ROMAIN RITZENTHALER

DEEEA/ETSE, Universitat Rovira I Virgili, 26 Avinguda dels Països Catalans,
Tarragona, 43007, Spain
romain.ritzenthaler@urv.cat

FRANÇOIS LIME

DEEEA/ETSE, Universitat Rovira I Virgili, 26 Avinguda dels Països Catalans,
Tarragona, 43007, Spain
francois.lime@urv.cat

This chapter presents some insights into the modeling of different Multi-Gate SOI MOSFET structures, and in particular Double-Gate MOSFETs (DG MOSFETs) and Tri-Gate MOSFETs (TGFETs). For long-channel case an electrostatic model can be developed from the solution of the 1D Poisson's equation (in the case of DG MOSFETs) and the 2D Poisson's equation in the section perpendicular to the channel (in the case of TGFETs). Allowing it to be incorporated in quasi-2D compact models. For short-channel devices a model can be derived from a 2D (in the case of DG MOSFETs) or a 3D (in the case of TGFETs) electrostatic analysis. The models were successfully compared with 2D and 3D TCAD simulations and, in some cases, experimental measurements. Short-channel effects, such as subthreshold slope degradation, threshold voltage roll-off and DIBL were accurately reproduced.

Keywords: compact modeling; Double-Gate MOSFETs; Tri-Gate MOSFETs; short-channel effects.

1. Introduction

This Chapter presents some recent developments in the field of modeling of Multi-Gate SOI MOSFETs (MuGFETs), and in particular Double-Gate MOSFETs (DG MOSFETs) and Tri-Gate MOSFETs (TGFETs), applicable to its related structures such as ΠFET and ΩFET structures by accounting for the full electrostatics of the devices, inclding the effect of the BOX, which brings an additional electrostatic coupling component from the back-gate. For long-channel devices, the electrostatic modeling can be developed from the solution of the 1D Poisson's equation (in the case of DG MOSFETs) and the 2D Poisson's equation in the section perpendicular to the channel (in the case of TGFETs). Assuming drift-diffusion transport, a compact drain current model for long-channel DG MOSFET is developed. The electrostatic model developed for TGFETs can be easily incorporated to quasi-2D compact drain current models by means of a correction in the threshold voltage. For short channel devices, the electrostatic modeling must be developed from an approximate solution of the 2D Poisson's equation (in the case of DG MOSFETs) and of the 3D Poisson's equation (in the case of TGFETs), since the electrostatics in the film or fin is affected by the longitudinal electric field arising from the drain. In all cases the Poisson's equation is solved in subthreshold, which is the regime where the electrostatic short-channel and interface coupling effects are dominant (in the above threshold regime these effects tend to vanish because of the screening of electric fields in the channel). The developed 3D electrostatic scheme for Tri-Gate MOSFETs can be incorporated to physically-based compact drain current models by means of corrections of the threshold voltage and subthreshold slope.

In section 2 we target the compact modeling of Double-Gate MOSFET (DG MOSFET) transistors. We present models for both symmetric and asymmetric long-channel DG MOSFETs (including in the latter the the case of Ultra-Thin Body SOI MOSFETs), developed from a charge control model derived from an approximate solution of the 1D Poisson's equation. Afterwards, from a quasi-2D analysis to solve Poisson's equation in subthreshold and in the saturated portion of the channel (in the saturation regime), we developed expressions to include the short-channel effects in the compact model for symmetric DG MOSFETs.

In Section 2, the threshold voltage modeling of long-channel Tri-Gate MOSFETs (TGFETs) is presented. The model was developed from an approximate solution of the 2D Poisson's equation in the plane perpendicular to the channel. The model can be extended to related structures, such as FinFETs, ΠFET and ΩFET. The back-interface parasitic activation is highlighted.

In Section 4, an analytical electrostatic potential model is developed for short-channel TGFETs. The analyical solution of the 3D potential is presented. The approximations made in order to develop an analytical subthreshold current model are explained and validated. The subthreshold current model was successfully compared with experimental data, in terms of subthreshold slope, threshold voltage roll-off and DIBL.

2. Compact modeling of Double-gate MOSFET (DG MOSFET) transistors

2.1. *Drain current and charge model for long channel DG MOSFETs*

In this work, fully analytical compact expressions for the charges and the drain current for each of the two channels of symmetrical Double Gate MOSFETs (DG MOS), Ultra-Thin Body (UTB) SOI and Asymmetric Double Gate MOSFETs (ADG MOS) with independent gate operation are presented. Most of the important physical parameters are given by the model, which makes it a good core model, easily adaptable to other devices and easily modifiable to incorporate additional effects. These analytical and compact expressions should also be useful for parameter extraction.

2.1.1. *Symmetric Double Gate MOSFET*

The device considered is shown in Fig. 1. The derivation presented in this part is based on [1] and [2]. We consider the long channel case in order to highlight the electrostatic coupling between the two gates by removing the influence of short channel effects.

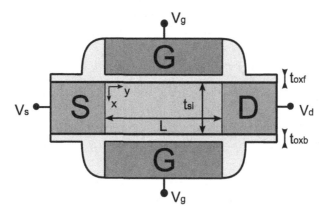

Fig. 1: Schematic representation of the symmetric DG MOS structure.

For long-channel devices, the gradual channel approximation could be used, which means that the problem reduces to solving the following 1D Poisson's equation:

$$\frac{\partial^2 \psi}{\partial x^2} = \frac{q n_i}{\varepsilon_{si}} \exp\left(\frac{\psi - V}{\beta}\right) \tag{1}$$

where $\beta = k_B T/q$, ε_{si} is the permittivity of the silicon, n_i the intrinsic concentration, and ψ the channel potential taken from the Fermi level in the source. V is the quasi-Fermi level position. Due to the symmetry of the device, we will solve Poisson's equation in the upper part of the silicon film. The total mobile charges or drain current will then be obtained multiplying by a factor of two.

The boundary conditions are:

$$\left.\frac{\partial \psi}{\partial x}\right|_{x=0} = -\frac{Q_g}{\varepsilon_{si}}; \left.\frac{\partial \psi}{\partial x}\right|_{x=x_0=t_{si}/2} = 0 \tag{2}$$

where Q_g is the gate charge.

Additionally, from Gauss's theorem:

$$Q_g = C_{ox}\left(V_g^* - \psi_s\right) \tag{3}$$

where $V_g^* = V_g - V_{fb}$, V_{fb} being the flat-band voltage and Ψ_s is the surface potential.

Integrating from the position of the middle of the channel x_0 , where the potential is Ψ_0, to an arbitrary point in the upper part of the silicon film,

$$\left(\frac{\partial \psi}{\partial x}\right)^2 - \left(\frac{\partial \psi}{\partial x}\bigg|_{x=x_0}\right)^2 = \frac{2qn_i}{\varepsilon_{si}}\beta\left(\exp\left(\frac{\psi-V}{\beta}\right) - \exp\left(\frac{\psi_0-V}{\beta}\right)\right) \tag{4}$$

Applying the boundary conditions, it can be expressed in the following form:

$$\left(\frac{\partial \psi}{\partial x}\right)^2 = \frac{2qn_i}{\varepsilon_{si}}\beta\exp\left(\frac{\psi-V}{\beta}\right) - \frac{C}{\varepsilon_{si}^2} \tag{5}$$

And, at the interface:

$$Q_g^2 + C = 2qn_i\beta\varepsilon_{si}\exp\left(\frac{\psi_s-V}{\beta}\right) \tag{6}$$

with $C = 2qn_i\beta\varepsilon_{si}\exp\left(\frac{\psi_0-V}{\beta}\right)$

C is a constant representing the electrostatic coupling between the two interfaces. A way to get rid of the unknown potential Ψ_0 in the middle of the silicon film, can be obtained as discussed in [2]: C only influences the behavior of the device in weak inversion, so it is convenient to approximate it to its value in this regime.

Integrating Eq. (5) from x_0 to the surface gives the following solution, expressed in terms of Q_g:

$$Q_g = \sqrt{C}\tan\left(\frac{\sqrt{C}}{4\beta C_{si}}\right) \tag{7}$$

Additionally, as in weak inversion C is supposed very small, a Taylor expansion can be done to give:

$$C \approx Q_g 4\beta C_{si} \tag{8}$$

Injecting in Eq.(6) and using the boundary conditions, the following charge control equation is found:

$$V_g - V_0 - V = \frac{Q_g}{C_{ox}} + \beta\ln\left(\frac{Q_g}{Q_0}\left(\frac{Q_g}{2Q_0}+1\right)\right) \tag{9}$$

with:

$$Q_0 = 2\beta C_{si}$$

$$V_0 = V_{fb} + \beta \ln\left(\frac{Q_0{}^2}{q n_i \varepsilon_{si} \beta}\right)$$

This equation is the charge control model for the front channel. To completely obtain the charge, Eq. (9) has to be solved numerically. However, an approximated solution can be obtained considering that it tends to two different limits in strong and weak inversion. Similarly to the method used in [3], we find for the strong inversion charge Q_{sif} :

$$Q_{si} = C_{ox}\left(V_g - V_{th} - V\right) \tag{10.a}$$

with $V_{th} = V_0 + \left(\beta \ln\left(\frac{Q_{si0}}{Q_0} + 1\right) + \beta \ln\left(\frac{Q_{si0}}{2Q_0} + 1\right)\right)$.

Q_{si0} is a modified strong inversion charge that tends to 0 in weak inversion, as explained in section 2.1.4.

For the weak inversion charge Q_{wif} we find:

$$Q_{wi} = Q_0\left(\sqrt{2\exp\left(\frac{V_g - V_0 - V}{\beta}\right) + 1} - 1\right) \tag{10.b}$$

These equations correspond to only half of the inversion charge as, due to the symmetry of the device, we solved Poisson's equation only in the upper part of the channel. It is also interesting to note that V_o is a threshold voltage consistent with the charge model.

In order to connect the two regimes together, it will be necessary to use interpolation functions (see section 2.1.4.).

2.1.2. *Asymmetric Double Gate MOSFET*

This section will generalize the approach given for the symmetrical DG MOS to the case of the asymmetrical one, with independent gate operation [4]. Such a model can also be used for Ultra-Thin Body SOI MOSFETs (UTB SOI).

The device considered is shown in Fig. 2. The derivation presented in this part is based on [5] and [6], but was extended to the case of ADG MOSFETs. We consider the long channel case in order to highlight the electrostatic coupling between the two gates by removing the influence of short channel effects.

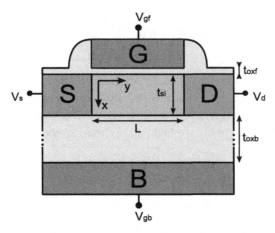

Fig. 2: Schematic representation of the asymmetric DG MOS structure.

In the following, the f index refers to the front gate and the b index to the back gate. Once again we will solve the same Poisson's equation as in the symmetric case but with the following boundary conditions that comes from by the conservation of the displacement vector through the gate dielectrics:

$$\left.\frac{\partial \psi}{\partial x}\right|_{x=0} = -\frac{Q_{gf}}{\varepsilon_{si}}; \left.\frac{\partial \psi}{\partial x}\right|_{x=t_{si}} = \frac{Q_{gb}}{\varepsilon_{si}} \tag{11}$$

where Q_{gf} and Q_{gb} are the charge densities at the front and back gate, respectively, as labeled in Fig. 3.

In addition, from Gauss's law, the following relations must hold:

$$Q_{gf} = C_{oxf}\left(V_{gf}^{*} - \psi_{sf}\right); Q_{gb} = C_{oxb}\left(V_{gb}^{*} - \psi_{sb}\right) \tag{12}$$

using the notations C_{oxf} and C_{oxb} for the front and back gate oxide capacitance, $V_{gf}^{*} = V_{gf} - V_{fbf}$, V_{fbf} being the front flat-band voltage and V_{gf} the front gate to source bias. Ψ_{sf} and Ψ_{sb} are the surface potentials at the front and back interfaces. Eq. (12) is also valid in the presence of volume inversion, as no charge-sheet approximation has been made. Indeed, these equations give the conservation of the charge between the gate and the S_i film, so it does not define the mobile charge distribution in the channel [6].

In order to solve the problem, the front and back channels need to be decoupled. This is done by making a decomposition of the charges in the structure as shown in Fig. 3. In this way, the channels are coupled together by a coupling charge Q_{cp} [7]. Q_{cp} can be viewed as the charge accumulated at the gate when the silicon film is depleted. From this point of view, it is then almost equivalent to the depletion charge of the bulk MOSFET. Assuming a fully depleted body, the coupling charge can be written:

$$Q_{cp} = C_{si}(\psi_{sf} - \psi_{sb}) \tag{13}$$

The two surface potentials ψ_{sf} and ψ_{sb} will vary in weak inversion and will be pinned at their threshold value in strong inversion. This means that it is only necessary to evaluate ψ_{sf}-ψ_{sb} at threshold condition and in weak inversion. In strong inversion, C_{si} will be decoupled from C_{oxf} and/or C_{oxb}. From Fig. 3 an expression of the coupling charge valid in weak and strong inversion can be derived:

$$Q_{cp} = C_{si}\left(V_{gf}^* - V_{gb}^* - \frac{Q_{gf}}{C_{oxf}} + \frac{Q_{gb}}{C_{oxb}}\right) = C_{eq}\left(\left|V_{gf}^{dep} - V_{gb}^{dep}\right|\right) \tag{14}$$

with $C_{eq} = \left(C_{si}^{-1} + C_{oxf}^{-1} + C_{oxb}^{-1}\right)^{-1}$ and $C_{si} = \dfrac{\varepsilon_{si}}{t_{si}}$.

$$V_{gf}^{dep} = V_{gf}^* - \frac{Q_{if}}{C_{oxf}}, \quad V_{gb}^{dep} = V_{gb}^* - \frac{Q_{ib}}{C_{oxb}}.$$

Q_{if} and Q_{ib} are the absolute values of the volume inversion charges respectively controlled by the front and the back gate. It is important to note that V_{gf}^{dep} is equal to V_{gf}^* in depletion and weak inversion and saturates to the threshold voltage minus the flat-band voltage in strong inversion. V_{gf}^{dep} should also saturate to a constant value in the accumulation regime, as it will be explained in the next sections.

The absolute value was added because the value of Q_0 should not change when permuting the front and back interfaces, and because Q_{cp}, Q_{if} and Q_{ib} have been defined so that they exhibit positive values.

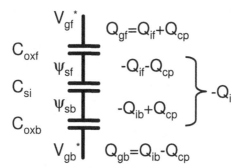

Fig. 3: Charges in absolute value and potentials in the weak inversion regime at various points in the asymmetrical structure.

In order to obtain an expression for the mobile charge, Eq. (1) should be integrated from the front to the back interface. Similarly to the solution derived in [8] it was decomposed as the sum of two integrals: the first integral corresponds to the front channel and its limits run from 0 to an arbitrary position x_0. The other integral from x_0 to t_{si} is for the back channel:

$$\int_0^{x_0}\frac{\partial^2\psi}{\partial x^2}dx + \int_{x_0}^{t_{si}}\frac{\partial^2\psi}{\partial x^2}dx = \int_0^{x_0}\frac{qn_i}{\varepsilon_{si}}\exp\left(\frac{\psi-V}{\beta}\right)dx$$

$$+ \int_{x_0}^{t_{si}}\frac{qn_i}{\varepsilon_{si}}\exp\left(\frac{\psi-V}{\beta}\right)$$

(15)

x_0 is chosen so that it corresponds to the position where the electric field is $-Q_{cp}/\varepsilon_{si}$ and the potential at this point is named ψ_0. In order to share the mobile charge between the two interfaces, we can make the assumption that the two equations for each integral can be solved independently. From here on we will only give the derivation of the front inversion charge, Q_{if}, as the derivation for the back channel is mostly identical and the final results for the back interface are the same with a permutation of the indices f and b.

For the front interface, it yields:

$$\left(\frac{\partial\psi}{\partial x}\bigg|_{x=0}\right)^2 - \left(\frac{\partial\psi}{\partial x}\bigg|_{x=x_0}\right)^2 =$$

$$\frac{2qn_i}{\varepsilon_{si}}\beta\left(\exp\left(\frac{\psi_{sf}-V}{\beta}\right)-\exp\left(\frac{\psi_0-V}{\beta}\right)\right)$$

(16)

The boundary condition corresponding to ψ_0 is:

$$\frac{\partial\psi}{\partial x}\bigg|_{x=x_0} = \frac{Q_{cp}}{\varepsilon_{si}}$$

(17)

When the back interface is in depletion, then $Q_{ib}=0$ and $x_0 = t_{si}$ corresponds to the position of the back interface. When the back inversion regime starts, x_0 is moving outside the channel. Applying the boundary conditions to Eq. (16), it yields:

$$\psi_{sf} - V = \beta\ln\left(\frac{Q_{gf}^2 + C}{2\beta qn_i\varepsilon_{si}}\right)$$

(18)

where $C = 2\beta qn_i\varepsilon_{si}\exp\left(\frac{\psi_0-V}{\beta}\right) - Q_{cp}^2$

As in the symmetric case, C represents the electrostatic coupling between the two interfaces [9] but in that case we had $Q_{cp}=0$. Furthermore it can be seen from Eq. (16) that the exponential term in C only has to be taken into account when Ψ_s is close to Ψ_0, which corresponds to the symmetrical case. This means that we can approximate its value to the one of the symmetrical case, that is to say $4\beta C_{si}Q_{if}$. This also ensures that the asymmetrical model will also be valid in the symmetrical case. Expressing the surface potential with the help of the boundary condition in Eq. (18) then yields:

$$V_{gf}^* - \frac{Q_{gf}}{C_{oxf}} - V = \beta \ln\left(\frac{(Q_{if}+2Q_0)Q_{if}}{2\beta qn_i\varepsilon_{si}}\right) \qquad (19)$$

where the coupling charge is modified as follows:

$$Q_0 = Q_{cp} + 2\beta C_{si} \qquad (20)$$

Next we express Q_{gf} as a function of Q_i:

$$Q_{gf} = C_{oxf}\left(V_{gf}^* - \psi_{sf}\right) = C_{oxf}\left(\frac{Q_{gb}}{C_{oxb}} + 2V_{dm} - (\psi_{sf} - \psi_{sb})\right) \qquad (21)$$

with $2V_{dm} = V_{gf}^* - V_{gb}^*$.

Simplifying Eq. (21) using the fact that $Q_i = Q_{gf} + Q_{gb} = Q_{if} + Q_{ib}$ yields:

$$Q_{gf} = Q_{if} + Q_{cp} = \frac{\gamma_f C_{oxf}}{1+\gamma_f}\left(\frac{Q_{if}}{\gamma_f C_{oxf}} + \frac{Q_{ib}}{C_{oxb}} + 2V_{dm}\right) \qquad (22)$$

where $\gamma_f = \dfrac{C_{oxb}}{C_{oxf}}\dfrac{C_{si}}{C_{oxb}+C_{si}}$

is the coupling factor for the front interface.

Replacing Eq.(22) into Eq. (19), it is found:

$$V_{gf} - V_{0f} - (1+\gamma_f)V =$$

$$\frac{Q_{if}}{C_{oxf}} + (1+\gamma_f)\beta \ln\left(\frac{Q_{if}}{Q_0}\left(\frac{Q_{if}}{2Q_0}+1\right)\right) \qquad (23)$$

with

$$V_{0f} = V_{fbf} - \gamma_f V_{gb}^{dep} + (1+\gamma_f)\beta \ln\left(\frac{Q_0^2}{\beta qn_i\varepsilon_{si}}\right)$$

This equation is the charge control model for the front channel. To completely obtain the charge, Eq. (23) has to be solved numerically. However, an approximated solution can be obtained considering that it tends to two different limits in strong and weak inversion. Similarly to the method used in [3], we find for the strong inversion charge Q_{sif}:

$$Q_{sif} = C_{oxf}\left(V_{gf} - V_{thf} - (1 + \gamma_f)V\right) \qquad (24.a)$$

with: $V_{thf} = V_{0f} + \left(1 + \gamma_f\right)\left(\beta \ln\left(\dfrac{Q_{si0}}{Q_0} + 1\right) + \beta \ln\left(\dfrac{Q_{si0}}{2Q_0} + 1\right)\right)$.

Q_{si0} is a modified strong inversion charge that tends to 0 in weak inversion, as explained in section 2.1.4.

For the weak inversion charge Q_{wif} we find:

$$Q_{wif} = Q_0\left(\sqrt{2\exp\left(\dfrac{V_{gf} - V_{0f} - (1+\gamma_f)V}{(1+\gamma_f)\beta}\right) + 1} - 1\right) \qquad (24.b)$$

From these equations it is also interesting to note that $(1+\gamma_f)$ is equivalent to a body factor and that V_{of} is a threshold voltage consistent with the charge model. The same equations are obtained for the back interface permuting the indices f and b.

In order to connect the two regimes together, it will be necessary to use interpolation functions (see section 2.1.4.   ).

To solve Eq. (24), we need an expression for V_{gb}^{dep} and V_{gf}^{dep}. These are needed to evaluate Q_0 , V_{0f} and V_{0b}. The threshold voltages can be rewritten in the approximated following form, with $V_{of}^* = V_{0f} - V_{fbf}$:

$$V_{0f}^* = -\gamma_f V_{gb}^{dep} + \left(1 + \gamma_f\right)V_t^{lim} \qquad (25.a)$$

$$V_{0b}^* = -\gamma_b V_{gf}^{dep} + \left(1 + \gamma_b\right)V_t^{lim} \qquad (25.b)$$

where V_t^{lim} is the minimum value of V_{of}^* and V_{ob}^* which is obtained when the opposite interface is in strong inversion. It is obtained by replacing V_{gb}^{dep} by V_{ob}^* in Eq. (25.a). In this model the found value is:

$$V_t^{lim} = \beta \ln\left(\dfrac{\left(C_{eq}\left|V_{gf}^* - V_{gb}^*\right| + 2\beta C_{si}\right)^2}{\beta q n_i \varepsilon_{si}}\right) \qquad (26)$$

but it should also be considered as a fitting parameter. This is corroborated by the fact that the relationships for V_{0f} and V_{0b} are similar to those of Lim and Fossum model for the threshold voltage of Fully Depleted SOI MOSFETs [10], where V_t^{lim} is indeed the value of the surface potential at threshold, that is to say half of the Si band-gap for an undoped device. This is the value we retained for the model. According to the Lim and Fossum model, the saturation of the threshold voltages in accumulation occurs at:

$$V_{gf}^* = V_{af} = -\frac{C_{si}}{C_{oxf}} V_t^{lim} \tag{27}$$

So the final expression for V_{gf}^{dep} (and similarly for V_{gb}^{dep}) is given by the following piecewise definition:

$$
\begin{aligned}
V_{gf}^{dep} &= V_{af} \quad for \quad V_{gf}^* \leq V_{af} \\
V_{gf}^{dep} &= V_{gf}^* \quad for \quad V_{af} < V_{gf}^* < V_{0f}^* \\
V_{gf}^{dep} &= V_{0f}^* \quad for \quad V_{gf}^* \geq V_{0f}^*
\end{aligned}
\tag{28}
$$

In the next section, mathematical functions are given to replace the conditional statements. The main advantage of these functions is to remove the necessity of having to calculate V_{gb}^{dep} in order to evaluate V_{gf}^{dep}, thus making the model explicit. Such functions are also convenient to avoid discontinuities and speed up computation.

Once the mobile charge is known, it is interesting to see that an expression for the surface potentials as a function of the mobile charge can be easily derived from Eq. (18) and Eq. (23).

2.1.3. *Smoothing functions for Q_0*

The saturation of the threshold voltage to V_{af} when the opposite gate is turned into accumulation (V_{af}) can be obtained by replacing V_{gf}^* with the following expression:

$$V_{gf}^a = V_{gf}^* + \left(1 + \gamma_f\right)\beta \ln\left(1 + \exp\left(\frac{V_{af} - V_{gf}^*}{\beta\left(1 + \gamma_f\right)}\right)\right) \tag{29}$$

The saturation of the threshold voltages toward V_t^{lim} in Eq. (25) can be conveniently represented by the following mathematical function and Eq. (25) becomes:

$$V_{0f}^* = V_t^{lim}\left(1 + \gamma_f \frac{\ln\left(1 + \exp\left(A\left(1 - \frac{V_{gb}^a}{V_t^{lim}}\right)\right)\right)}{\ln(1 + \exp(A))}\right) \tag{30}$$

Similarly, to avoid discontinuities, the saturation of V_{gf}^{dep} and V_{gb}^{dep} to their respective threshold voltage was obtained with the following function:

$$V_{gf}^{dep} = V_{of}^{*}\left(1 - \frac{\ln\left(1 + \exp\left(A\left(1 - \frac{V_{gf}^{a}}{V_{0f}^{*}}\right)\right)\right)}{\ln(1 + \exp(A))}\right) \tag{31}$$

with A being a smoothing parameter (A=8 was found to be an adequate value).

2.1.4. Interpolation functions

The equations given in this section as well as the next one apply to both the symmetrical and asymmetrical case. Indeed, it can be seen that the charge control equations for both cases are the same, except for the values of Q_0 and V_0, and γ that is equal to 0 in the symmetrical case. In other words the symmetrical case is a special case of the asymmetrical one, and this is taken into account in the asymmetrical model.

In order to connect (24.a) and (24.b) together, we start by increasing the range of validity of (24.a) by making it tend smoothly to 0 in weak inversion. This is achieved rewriting it as:

$$Q_{sif(b)} = C_{oxf(b)}\left(\eta_1 + \eta_2\gamma_{f(b)}\right)\beta$$
$$\ln\left(1 + \exp\left(\frac{V_{gf(b)} - V_{thf(b)} - (1 + \gamma_{f(b)})V}{\left(\eta_1 + \eta_2\gamma_{f(b)}\right)\beta}\right)\right) \tag{32}$$

with,

$$Q_{sif(b)0} = C_{oxf(b)}\left(\eta_1 + \eta_2\gamma_{f(b)}\right)\beta\ln\left(1 + \exp\left(\frac{V_{gf(b)} - V_{0f(b)} - (1 + \gamma_{f(b)})V}{\left(\eta_1 + \eta_2\gamma_{f(b)}\right)\beta}\right)\right),$$

where η_1 controls the abruptness of the transition for the front gate operation, while η_2 controls it for an operation from the back gate, that is to say when the transistor is turned on by variations in V_{th}. Typical values for η_1 and η_2 are 2.5 and 3.5, respectively.

The total charge $Q_{if(b)}$ for the front (back) channel is then expressed as:

$$Q_{if(b)} = \left(\frac{Q_{sif(b)}^{1/\eta_3}}{1 + \left(\frac{Q_{sif(b)}}{Q_{wif(b)}} \right)^{\frac{1}{\eta_3}}} \right)^{\eta_3} \tag{33}$$

where the value for η_3 is typically between 1 and 1.5. $Q_{wif(b)}$ is given by Eq. (24.b).

This method allows an accurate fit of the moderate inversion region with the adjustment of the three parameters described above.

2.1.5. *Drain current model*

As in [5,6] the drain current is expressed as:

$$I_{ds} = \frac{W}{L} \mu_{eff} \int_0^{V_{ds}} Q_i(V) dV \tag{34}$$

where μ_{eff} is the effective mobility, W is the device width and L is the channel length.

From the charge control model Eq. (23), it is obtained:

$$dV = -\frac{dQ_i}{(1+\gamma)C_{ox}} - \beta \left(\frac{dQ_i}{Q_i} + \frac{dQ_i}{Q_i + 2Q_0} \right) \tag{35}$$

Integrating Eq.(34) using Eq.(35) between Q_s and Q_d ($Q_i = Q_s$ at the source end and $Q_i = Q_d$ at the drain), the drain current I_{ds} for either the front or back channel can be expressed in terms of carrier charge densities as:

$$I_{ds}(Q_s, Q_d, \gamma, C_{ox}) =$$
$$\frac{W}{L} \mu_{eff} \left[\frac{Q_s^2 - Q_d^2}{2(1+\gamma)C_{ox}} + 2\beta(Q_s - Q_d) + 2\beta Q_0 \ln\left(\frac{Q_d + 2Q_0}{Q_s + 2Q_0} \right) \right] \tag{36}$$

The current relative to the front and back channels I_f and I_b are then obtained from Eq. (36) for the front and back channel, respectively,

$$I_f = I_{ds}(Q_{ifs}, Q_{ifd}, \gamma_f, C_{oxf})$$
$$I_b = I_{ds}(Q_{ibs}, Q_{ibd}, \gamma_b, C_{oxb}) \tag{37}$$

where Q_{fis} and Q_{fid} are respectively obtained from Eq. (33) with V=0 and V=V_{ds}. Finally, the sum of the front and back channel currents gives the total source to drain current of the device:

$$I = I_f + I_b \tag{38}$$

For the symmetrical case, I_f and I_b are equal, so we recover the factor 2 that was missing.

2.1.6. Results and discussion

In summary, the model consists in the following procedure: first, evaluate V_{gb}^{dep} and V_{gf}^{dep} using Eq. (31), then calculate Q_0 using Eq. (20), which allows calculating the threshold voltages V_{0f} and V_{0b} given in Eq. (23). The mobile charges for each of the two channels are then obtained using Eq. (32), Eq. (24.b) and Eq. (33) and the contribution of this channel to the drain current is given by Eq. (36) and Eq. (37). Finally, the total drain current is given by Eq. (38).

Compared to previously published models like [8], [11,14], the presented model has the advantage of being fully analytical, as these models require numerical or iterating solving at some point. The result is a unique solution for all the cases, with simpler equations; this makes the model fully analytic and much easier to implement. However the precision is smaller if the approximated solution to Eq. (23) is used, which is necessary to obtain analytical and compact solutions for the model. The analytical solutions in Eq. (24) present some similarities with [7] but the parameters V_0 and Q_0 are completely different. This comes from the fact that in our model, these equations were rigorously obtained from Poisson's equation, while in [7] these are a starting hypothesis of the derivation.

Thanks to the physical basis of the model, most physical quantities are readily given, such as the threshold voltage V_0, the front and back channel currents, the slope factor, the inversion charge. In addition, the values of the input parameters (mobility, flat-band voltages, device dimensions,...) were found to be very close to the ones used in the simulations, confirming the physical consistency of the model. The main parameter discrepancy could come from the doping as the model assumes an intrinsic body. However this could be settle in a first approximation by a shift in the flat-band voltages. Because of this doping discrepancy, the accumulation voltage V_a should be considered a fitting parameter but its influence on the model is not determinant. In addition, because of the lack of majority carrier sources in the device, the accumulation is unlikely to occur, so a much higher value for V_a than the theoretical one is expected. Moreover, this model does not pretend to provide an accurate description of the device in the accumulation regime.

All the TCAD simulations were done with default parameters. This means that they are not representative of a real device, especially for the mobility which should be much lower and V_g dependent. The interest of doing so is that the mobility is often considered as a complex fitting parameter, so by using the same constant value than in the simulation, we ensure that the model presents the correct electrostatic behavior, as it allows distinguishing the electrostatic behavior and the effects that come from the transport. All the TCAD simulations were done with a p type doping of $1 \times 10^{15} \text{cm}^{-3}$. Also, for the sake of simplicity, we used in the simulations the same flat band voltages for both gates, and the same mobilities for both interfaces.

As seen before, in the presented model, the drain current is split in two contributions: a front channel current and a back channel current. These are not obtained with the charge sheet approximations as volume inversion is taken into account for each of these contributions but, from the hypothesis that have been made, the charge distribution of each contribution is higher at their respective interface. In fact, the model is partitioning the mobile charge between a front gate controlled and a back gate controlled one, thus giving rise to a front and a back channel current, as derived by the model. This partitioning of the mobile charge is shown in Fig. 4, where can be seen the currents from the front and back channels as well as the total current. In this UTB SOI FET case, the back channel is activated by a lowering of its threshold voltage due to the applied bias at the front gate. The threshold voltages for the two channels are, of course, different. The saturation of the curve occurs when the front channel inverts itself, thus screening the influence of the front gate. This saturation is in fact an approximation, but as it is only affecting the back channel current in a region where it is always smaller than the front channel current, the impact on the total current is negligible. For the same reason, the back channel current also presents a plateau at low gate voltages, when the front channel is in the accumulation regime. Another thing to see in this figure is that for the same flat-band voltage, when the back oxide is thicker than the front one, the resulting back channel current is lower than the front one, especially in strong inversion, but that is less true in weak inversion. This means that the assumption of neglecting the charges at the back interface that is usually made for UTB SOI is only a rough approximation in this regime.

It is well known that an applied bias at the back gate changes the threshold voltage of the device. In this model, this is represented as an earlier activation of the back channel for positive back gate biases, the effect on the front channel being much smaller. In fact the general threshold voltage of the device is then the lowest one of the two.

We can see in Fig. 5 that the model reproduces well the variations of the threshold voltage when changing the opposing gate bias, even when operating from the back gate. However, we can see that the $I_d(V_{gb})$ curves present a hump around 0 Volts which is not reproduced by the simulations but this hump is also observed in electrical measurements and is generally attributed to substrate depletion. Here, this effect is not taken into account and the hump was found to be due to a canceling of the interfaces coupling, as it corresponds to the minimum value of the coupling charge Q_0. Nevertheless the accordance is good enough considering that the device is not supposed to work in this bias configuration.

Fig. 5 shows that the model gives good results for asymmetrical gate operation. Furthermore, we checked (not shown) that the model is also valid for the limiting case of the symmetric DG MOSFET. In fact, the asymmetrical case converges to the symmetrical one. Indeed, due to different boundary conditions, the symmetrical and asymmetrical models are often different [9], but in this model, the unification of both cases is ensured with the addition of the term $2\beta C_{si}$ to the coupling charge Eq. (20).

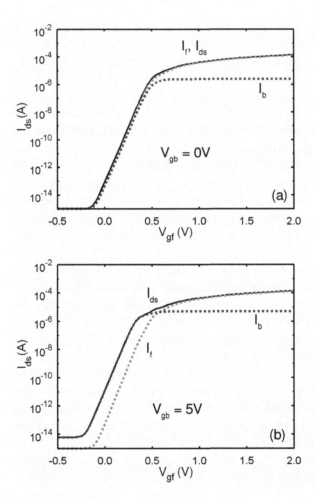

Fig. 4: $I_{ds}(V_{gf})$ characteristics at V_{ds} = 50mV for a) V_{gb} = 0V and b) V_{gb}=5V showing the contribution to the total current of the front and back channel currents I_f and I_b. The device dimensions are L=W= 1μm, t_{oxf}=1.65nm EOT, t_{oxb} =145nm, t_{si} =13nm.

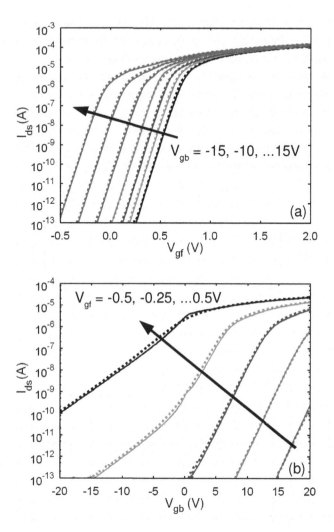

Fig. 5: Interface coupling characteristics at V_{ds} = 50mV (model = lines) compared with TCAD simulations (dashed lines) : a) $I_{ds}(V_{gf})$ curves for V_{gb} = -15, -10, ..., 15V, and b) $I_{ds}(V_{gb})$ curves for V_{gf} = -0.5, -0.25, ..., 0.5V. The device dimensions are L=W= 1μm, t_{oxf}=1.65nm EOT, t_{oxb} =145nm, t_{si} =13nm.

In Fig. 6 the model is compared with experimental measurements. The UTB devices were provided by CEA-LETI (Grenoble, France) and consist of a "wide" FinFET technology with a low doped Si body. The gate stack consists of a HfO_2/SiO_x bilayer dielectric with a TiN/W metal gate. More details on the process can be found in [15]. A very good agreement was obtained by only adjusting the mobility and flat-band voltages, so the model could be used for parameter extraction. The GIDL observed at negative gate biases is not taken into account by the model. For the case of V_{gb} = 0 considered here, we used the same mobility values for the front and back interfaces as the back channel current should be negligible in strong inversion, and errors in weak inversion should be compensated by a small shift in the flat-band voltages. The values we used are: $V_{fbf}=0.05V$ and $V_{fbb}=0.5V$ for the flat-band voltages, and $\mu_0=180cm^2/V/s$, $\theta_1=-0.1V^{-1}$ and $\theta_2=0.15V^{-2}$ for the low field mobility, and the first and second order mobility attenuation factors θ_1 and θ_2, the series resistance being included in θ_1 [1]. As volume inversion is generally only noticeable in weak inversion, the difference in mobility value between the volume of the S_i film and the surface should be taken into account by the gate voltage dependence of the mobility. Finally, this also shows that for this relatively thin silicon thickness value of $t_{si}=13nm$, it is still possible to accurately simulate the behavior the device with realistic parameters without taking quantum effects into account

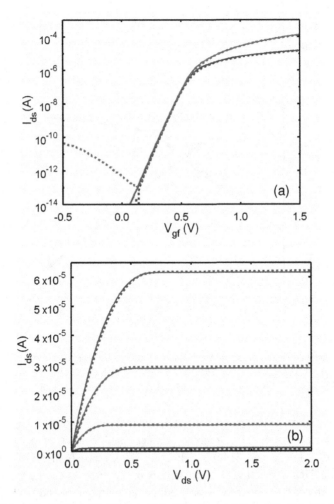

Fig. 6: a) $I_{ds}(V_{gs})$ at $V_{ds}=0.05V$ and 1.2V, and b) $I_{ds}(V_{ds})$ characteristics at $V_{gf}=0.6, 0.8, 1, 1.2V$ fitted on experimental curves with $V_{gb} = 0$. Solid lines are the model. Dashed lines are the measurements. The device dimensions are L=W= 10μm, $t_{oxf}=1.65nm$ EOT, $t_{oxb} =145nm$, $t_{si} =13nm$.

Summarizing, a compact model for Asymmetric Double-Gate MOSFETs with independent gate operation has been presented, with the following advantages: fully analytical and explicit derivation with no iterative solutions, accessible front and back gate charges, potentials and currents, unification of symmetric and asymmetric cases, physical solutions similar to classical MOS theory. The model has the following limitations: long channel model, undoped channel, the moderate inversion regime is interpolated from the weak and strong inversion solutions via three fitting parameters. Although we only presented the drain current model, analytical expressions for the capacitances can be obtained as explained in [5].

2.2. *Short channel model for symmetric DG MOSFET*

This part presents a way to model short channel effects in symmetric DG MOSFETs. The case of asymmetrical DG MOS is not presented because, to our knowledge, it still has not been done properly and short channel effects in ADG MOS are still not well understood.

2.2.1. *Velocity saturation*

One of the most used expressions for the saturation potential for electron, is [16]

$$V_{sat} = (V_{gs} - V_t)\frac{v_{sat}}{\frac{\mu_{eff}}{2L}(V_{gs} - V_t) + v_{sat}}$$

μ_{eff} being the effective mobility, v_{sat} the saturation velocity, and V_t the threshold voltage. To extend the validity of this relation to weak inversion, we replaced V_{gs}-V_t by $Q_s/(2C_{ox})$ [16], which is a term that tends to V_{gs}-V_t in strong inversion and to 0 in low inversion. Q_s being the charge at the source end of the channel. Thus,

$$V_{sat} = (-Q_s/2C_{ox})\frac{v_{sat}}{-Q_s\frac{\mu_{eff}}{4LC_{ox}} + v_{sat}} \tag{39}$$

However, with this equation, V_{sat} tends to 0 in weak inversion, whereas the theoretical value is $2k_bT/q$. To correct this, we propose to replace Q_s in Eq. (39) with Q_{seff} so that we have:

$$V_{sat} = (-Q_{seff}/2C_{ox})\frac{v_{sat}}{-Q_{seff}\frac{\mu_{eff}}{4LC_{ox}} + v_{sat}}$$

with

$$Q_{seff} = Q_s + 4\frac{k_bT}{q}C_{ox}\frac{v_{sat}}{v_{sat} - \frac{k_bT}{q}\frac{\mu_{eff}}{L}} \tag{40}$$

This way, Q_{seff} tends to Q_s in strong inversion, and to a value that gives the correct V_{sat} in weak inversion.

The usual approximated relation for the mobility dependence with the longitudinal electric field is:

$$\mu = \frac{\mu_{eff}}{1 + \frac{\mu_{eff}}{v_{sat}}\frac{V_{sat}}{L - \Delta L}} \tag{41}$$

where ΔL is the length of the saturation region. However, the results have been found to be far better with the more precise relation:

$$\mu = \frac{\mu_{eff}}{\left(1 + \left(\frac{\mu_{eff}}{v_{sat}}\frac{V_{sat}}{L - \Delta L}\right)^{n_m}\right)^{1/n_m}} \tag{42}$$

with $n_m=2$.

In fact, Eq. (41) is a usual approximation of Eq.(42).

If we replace the above expression in the equation of the drain current, we obtain :

$$I_{DS} = 2\frac{W\mu_{eff}}{L_e}$$

$$\left[2\frac{k_BT}{q}(Q_s - Q_d) + \frac{Q_s^2 - Q_d^2}{2C_{ox}} + 4\left(\frac{k_BT}{q}\right)^2 C_{Si}\log\left[\frac{Q_d + 2Q_0}{Q_s + 2Q_0}\right]\right] \tag{43}$$

In this way, I_{DS} is expressed in terms of μ_{eff}. Velocity saturation is taken into account in L_e, which can be considered as an effective gate length due to the effect of velocity saturation [17]:

$$L_e = (L - \Delta L)\left[1 + \left(\frac{\mu_{eff}V_{sat}}{v_{sat}(L - \Delta L)}\right)^{n_m}\right]^{\frac{1}{n_m}} \tag{44}$$

We found necessary to use the exact relation for longitudinal field dependence of the mobility, as the approximated relation was fairly inaccurate in reproducing our ATLAS drain current simulations.

For the model to be continuous during the transition to saturation regime, we need to introduce an effective drain voltage, that is equal to V_{ds} in the linear regime and that becomes continuously and progressively equal to V_{sat} in the saturation regime. A possible expression is [18]:

$$V_{deff} = V_{sat} - V_{sat} \frac{\ln\left[1+\exp\left(1-\frac{V_{ds}}{V_{sat}}\right)\right]}{\ln[1+\exp(A)]} \tag{45}$$

where A is a fitting parameter which defines the abruptness of the transition V_{ds} to V_{sat}. A value of A=6 worked well in our case.

2.2.2. *Series resistances*

For not too high series resistances R_s, its effect can be modeled to the first order, as a corrective term in the mobility expression:

$$\mu_{eff} = \left(\left(\mu_{eff}^{0}\right)^{-1} + \left(\mu_{Rs}\right)^{-1}\right)^{-1} \tag{46}$$

where μ_{eff} and μ_{eff}^{0} are respectively the effective mobility with and without series resistance, and

$$\mu_{Rs} = 2C_{ox}R_S W/(L-\Delta L)\left(Q_d/(2C_{ox})+0.5V_{eff}\right)$$

A possible well known expression for μ_{eff} is:

$$\mu_{eff}^{0} = \frac{\mu_0}{1+\theta_1 \dfrac{Q_s}{2C_{ox}} + \theta_2 \left(\dfrac{Q_s}{2C_{ox}}\right)^2} \tag{47}$$

where μ_0, θ_1 and θ_2 are respectively the low field mobility, and mobility attenuation coefficients of the first and second order, which can be viewed as fitting parameter.

2.2.3. *Channel length modulation*

In order to model the channel length modulation, we need to solve the Poisson equation in the saturation region. It can be written, considering a symmetric undoped double gate nMOS, as:

$$\frac{\partial^2 \phi}{\partial x^2} + \frac{\partial^2 \phi}{\partial y^2} = -\frac{-q\,n_e(x, y)}{\varepsilon_{si}}$$ (48)

where n_e is the mobile electron charge concentration.

We use the following boundary conditions:
The electric field cancels at the middle of the Si film, $y=t_{si}/2$.

$$\left.\frac{d\phi}{dy}\right|_{y=tsi/2} = 0$$

The displacement vector is continuous at the interface, $y=0$.

$$\varepsilon_{si}\left.\frac{d\phi}{dy}\right|_{y=0} = \varepsilon_{ox}\frac{\phi_S - V_{gs} + \Delta\phi}{t_{ox}}$$

$\phi_s=\phi(y=0)$ is the potential in Si at the interface and $\Delta\phi$ is the work function difference between the gate electrode and the intrinsic silicon.

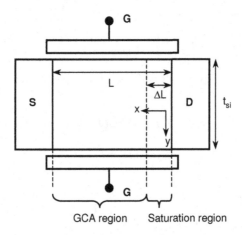

Fig. 7: Schematic representation of the DG MOSFET considered here.

Like in other works, we chose a power law as an approximation for the potential profile along axis y [19, 20, 21]. This gives simpler solutions to Poisson equation. However, unlike those works, we are not in threshold conditions here so we chose not to use a parabolic profile, but to consider the exponent as a parameter n whose value will be discussed in the next section:

$$\phi(x, y) = a + b(x)y + c(x)y^n$$ (49)

The parameters a, b and c are evaluated by applying the boundary conditions, which yields the following potential profile:

$$\varphi(x,y) = \varphi_S + \frac{\varepsilon_{ox}}{\varepsilon_{si}} \frac{\varphi_S - V_{gs} + V_{FB}}{t_{ox}} y$$

$$- \frac{\varepsilon_{ox}}{\varepsilon_{si}} \frac{\varphi_S - V_{gs} + V_{FB}}{n \cdot t_{ox}} \left(\frac{2}{t_{si}}\right)^{n-1} y^n \tag{50}$$

for $y \le t_{si}/2$

$$\phi(x,y) = \phi_S + \frac{\varepsilon_{ox}}{\varepsilon_{si}} \frac{\phi_s - V_{gs} + \Delta\varphi}{t_{ox}} (t_{si} - y) - \frac{\varepsilon_{ox}}{\varepsilon_{si}} \frac{\phi_s - V_{gs} + \Delta\varphi}{n \cdot t_{ox}} \left(\frac{2}{t_{si}}\right)^{n-1} (t_{si} - y)^n$$

for $y > t_{si}/2$

One should note that in Eq. (49), if n=1 we have a flat profile, while n=2 gives a parabolic profile.

Then integrating Eq. (48) over and using Eq. (50) gives :

$$\int_0^{tsi} \frac{\partial^2 \phi}{\partial x^2} dy = -\frac{Q_m}{\varepsilon_{si}} + 2 \frac{\varepsilon_{ox}}{\varepsilon_{si}} \frac{\phi_s - V_{gs} + V_{FB}}{t_{ox}} \tag{51}$$

where Q_m is the mobile charge in the saturated region integrated over the silicon thickness, which is negative for a nMOS.

Evaluating the left-hand side of Eq. (51) using Eq. (50) gives :

$$\frac{\partial^2 \phi_S}{\partial x^2} = -\frac{Q_m}{\lambda^2 2C_{ox}} + \frac{\phi_s - V_{gs} + \Delta\varphi}{\lambda^2} \tag{52}$$

where:

$$\lambda = \sqrt{\frac{\varepsilon_{si} t_{ox} t_{si}}{2\varepsilon_{ox}} + \frac{t_{si}^2}{8}\left(1 - \frac{2}{n(n+1)}\right)} = \frac{t_{si}}{2}\sqrt{\frac{1}{2} + \frac{1}{2r} - \frac{1}{n(n+1)}} \tag{53}$$

λ is a characteristic length that depends only on the device structure and the chosen potential profile. λ can be rewritten in terms of $r = C_{ox}/4C_{si}$, which is a structural parameter more relevant of the physical behavior of the device, C_{ox} and C_{si} being respectively the gate dielectric and the silicon film capacitance.

If we consider that Q_m is constant along the channel, then we can easily solve Eq. (52). Indeed Eq. (52) becomes:

$$\frac{\partial^2 \varphi}{\partial x^2} - \frac{\varphi}{\lambda^2} = 0 \tag{54}$$

with $\varphi = \phi_S - V_{gs} + \Delta\varphi - \dfrac{Q_m}{2C_{ox}}$.

Eq. (54) is solved with the following boundary conditions, considering the origin of the x axis at the drain (see Fig. 7):

$$\varphi(x = -\Delta L) = \varphi(\phi_S = V_{deff} + \phi_b) = \varphi_{sat} ,$$

$$\frac{d\varphi}{dx}\bigg|_{x=-\Delta L} = \frac{k v_{sat}}{\mu}$$

ΔL, v_{sat}, V_{sat} and μ being respectively the length of the saturation region, the saturation velocity, the saturation voltage and the effective mobility. ϕ_b is the surface potential at the source and at threshold condition, such that $\Delta\varphi+\phi_b=V_t$. Here, k=2 as we are considering nMOSFETs. k=1 for a pMOS. This gives the following solution for φ:

$$\varphi(x) = \varphi_{sat} \cosh\left(\frac{\Delta L + x}{\lambda}\right) + \frac{k v_{sat}}{\mu} \lambda \sinh\left(\frac{\Delta L + x}{\lambda}\right) \tag{55}$$

ΔL is then obtained from $\varphi_d = \varphi(x = 0) = \varphi(\phi_S = V_{ds} + \phi_b)$:

$$\Delta L = \lambda \ln\left(\frac{\varphi_d + \sqrt{\varphi_d^{\,2} - \varphi_{sat}^{\,2} + \left(\dfrac{k v_{sat}}{\mu} \lambda\right)^2}}{\varphi_{sat} + \dfrac{k v_{sat}}{\mu} \lambda}\right) \tag{56}$$

with:

$$\varphi_{sat} = V_{deff} - V_{gs} + V_t - \frac{Q_m}{2C_{ox}} \approx V_{deff} + \frac{Q_s + Q_m}{4C_{ox}} - \frac{Q_m}{2C_{ox}}$$

$$\varphi_d = V_{ds} - V_{gs} + V_t - \frac{Q_m}{2C_{ox}} \approx V_{ds} + \frac{Q_s + Q_m}{4C_{ox}} - \frac{Q_m}{2C_{ox}}$$

These expressions are equivalent to what has been obtained for bulk MOSFETs, differing only by the expressions for the parameters φ_d, φ_{sat} and λ [22]. Here $Q_m =$ $-Q(V_{gs}, V_{deff})$, as given by equations (33) and (45), and $Q_s = -Q(V_{gs},0)$.

It is possible to obtain an expression similar to Eq. (56) for the potential in the middle of the Si film. However with this approach, it is necessary to know an expression of the saturation potential in the middle of the film, which is less convenient since the well-known equation (39) that we used here is for the surface potential.

2.2.4. DIBL effect

The DIBL effect can be modeled by solving the Poisson equation in the same way as for the channel length modulation. As the DIBL takes place in weak inversion, and therefore concerns mainly a conduction in the depth of the silicon film, it is more rigorous here to consider the potential at the center ϕ_c, located at $y=t_{si}/2$, similarly to what has been done in [21]. However, we have to say that we observed nearly no differences by calculating the DIBL from the surface potential. ϕ_c is obtained from Eq. (50), as a function of ϕ_s. We then have to replace ϕ_s in Eq. (52). We finally obtained the same equation as Eq. (54) but the boundaries conditions and the variable φ are different:

$$\varphi(x=0)=\varphi(\phi_c = V_{deff} +V_{bi})=\varphi_d$$

$$\text{and } \varphi(x=-L)=\varphi(\phi_c = 0+V_{bi})=\varphi_s \text{ at the source.}$$

V_{bi} being the source and drain junction built-in voltage. As we will see below, we assume here that V_{sat} at the center of the silicon film is approximately the same as the one at the surface.

The variable φ is this time equal to

$$\varphi = \phi_c - V_{gs} +\Delta\varphi-\left(1+\frac{C_{ox}}{2C_{si}}\left(1-\frac{1}{n}\right)\right)((Q_s +Q_d)/2)\frac{1}{2C_{ox}}$$

Q_s and Q_d being the mobile charge at the source and drain, respectively. We chose here to approximate the mobile charge to the average charge in the channel, because we need a constant charge for the integration of Eq. (54) along the channel.

Then we find the following expression for φ [20]:

$$\varphi(x)=\frac{\varphi_s \sinh\left(\frac{L-x}{\lambda_{DIBL}}\right)+\varphi_d \sinh\left(\frac{x}{\lambda_{DIBL}}\right)}{\sinh\left(\frac{L}{\lambda_{DIBL}}\right)} \tag{57}$$

where λ_{DIBL} is the characteristic length for the DIBL, that is to say Eq. (53) with n=2, which correspond to the potential profile usually taken in weak inversion.

The minimal potential for φ is:

$$\varphi_{\min} = 2\sqrt{\varphi_s \varphi_d} \, \exp\left(\frac{-L}{2\lambda}\right) \qquad (58)$$

This quantity, equal to 0 for long channel devices [20], can be considered as the barrier potential drop due to the DIBL effect and must not be confounded with the surface potential which is taken into account in the V_t parameter in the CLM section. It is then introduced into the calculation of the charge Q_{wi} in Eq. (33) and Eq. (10.b), by replacing V_0 with V_0-$\phi_{\min}$, for example.

2.2.5. *Results and discussion*

Usually, around the threshold voltage, the value of the n parameter is taken as 2 in the gradual channel approximation (GCA) region, i.e. a parabolic potential profile. However this is not true in our case in the saturation region. In fact, Silvaco-Atlas simulations give a somewhat parabolic profile in the GCA region that becomes flat near the boundary between the saturated region and the GCA region. This flat potential profile corresponds to n=1 in Eq. (53). In the saturated region, ATLAS gives a potential profile that presents a very small curvature that can be fitted with a n value slightly inferior to 1, that depends on the device geometry.

Therefore, it is important to note that in our model, we consider two different constant values of n for the GCA and saturated region, respectively. For the evaluation of the channel length modulation, we assume that n=1 in the saturated region is a good approximation for the devices considered here. Furthermore, we observed that the value of n has little influence on the channel length modulation. However, for the DIBL calculation, i.e in the expressions of ϕ_s, ϕ_d and λ_{DIBL}, the value of n is taken as 2, because DIBL mainly occurs in weak inversion.

The model is for undoped devices, but a channel doping of 10^{15}cm^{-3} was used for the simulations. Anyway, a model for an undoped transistor can be applied to a lightly doped one by adjusting the value of the flat band voltage [23]. The doping of the source and drain region is 6×10^{20}cm^{-3}. The gate and source workfunctions are 4.73 and 4.1eV, respectively.

Concerning the calculation of the DIBL, the exact value of V_{bi} is difficult to determine in a double gate MOSFET because the silicon film is floating. Therefore, V_{bi} was taken as a fitting parameter for the DIBL. In Fig. 4, we present a fitting of the I_d-V_d and I_d-V_g curves given by Silvaco simulations. In order to reduce the number of fitting parameters and facilitate the comparison with ATLAS, we chose a constant mobility with V_g in the simulations of Fig. 8, but we used in the model the mobility extracted from

ATLAS curves. We can see that the overall agreement for the current is very good, especially for the DIBL, which gave the correct qualitative and quantitative behavior below the threshold, without the use of any fitting parameters. The correct quantitative behavior was obtained with the right V_{bi}.= 0.57V. Concerning the dynamic conductance G_d, the agreement between ATLAS and the model is reasonably good below 1.5V in Fig. 4c. Beyond 1.5V, the model overestimates the slope of G_d, but nanoscale DG MOSFETs are not supposed to work in the high V_{gs} regime. This is due to the fact that the model seems to underestimate slightly the channel length modulation when short channel effects are very strong. The short channel model is further verified in Fig. 9, which shows excellent agreement with ATLAS when the channel length is varied, illustrating the accuracy of the DIBL modeling. Finally, we can see that the subthreshold slope illustrated in Fig. 10 is quite close to the one given by the simulations. Although we presented results for only one device geometry, we verified that the model is geometrically scalable, at least for not too small gate lengths.

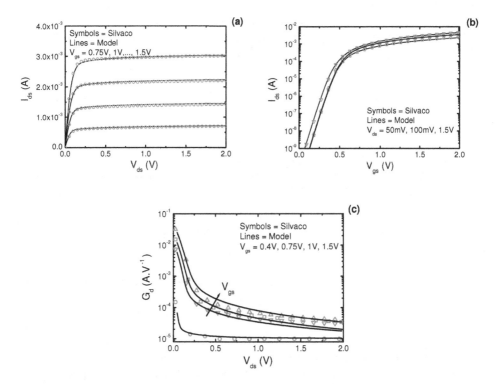

Fig. 8: Comparison of the model with Silvaco simulations, for a DG MOS with t_{ox}=2nm, t_{si}=15nm and L= 50nm.

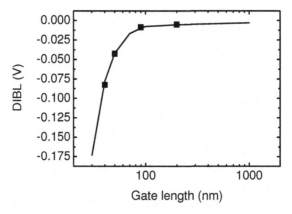

Fig. 9: DIBL defined in weak inversion as the V_{gs} offset between $I_{ds}@V_{ds}=1V$ and $I_{ds}@V_{ds}=50mV$, as given by the model (lines) and ATLAS simulations (symbols). $t_{si}=15nm$, $t_{ox}=2nm$.

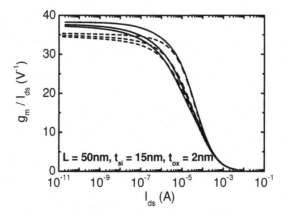

Fig. 10: gm/I_{ds} curves at various $V_{ds}=50mV$, 1V, 1.5V, as given by the model (lines) and ATLAS simulations (dashed lines).

3. Analytical modeling of Triple-gate, Π-gateFETs, and Ω-gateFETs devices

3.1. *Introduction*

3.1.1. *Structure*

The downscaling of CMOS transistors below the 32 nm node is challenging [24], and the control of leakage currents needs such thin gate oxides and highly doped channels that a process could be hard to find. Therefore, new architectures called Multiple-Gate FETs (or MuGFET transistors) made on Silicon On Insulator (SOI) or bulk wafers have been proposed for the next technological nodes (in 2006, the world records in term of miniaturization (L_G = 5 nm for NMOS [25] and PMOS [26]) were held by MuGFET transistors). Multiple-gate architectures have been introduced in the 1980's (Fig. 11) and combine a number of advantages: the electrostatic coupling between the gates allows to control better the channel and therefore to reduce the short channel effects and the leakage in the transistor in the off-state; decreasing the transverse electric field, the carrier's mobility will also be improved. Additionally, the on-state current will be increased if the conduction channel surface is increased compared to a planar transistor. Regarding the process, a small variation compared to a classical CMOS process is pursued in order to minimize the technological cost related to the transistor architecture change.

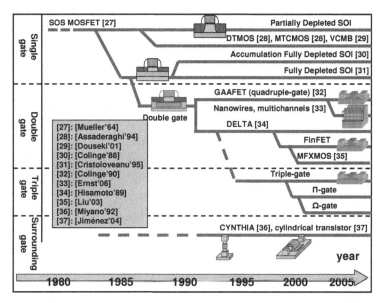

Fig. 11: Historical evolution of multiple-gate transistors. The devices not described in this chapter are reported in the reference list. Figure adapted from [38].

In a FinFET transistor, the top channel is not active (at least not significantly compared to other channels) because of the nitride hard mask [39]. A Triple-gate FET (or TGFET) is a transistor where the hard mask has been etched before the gate stack deposition (Fig. 12, [40]). It is therefore a transistor with two vertical conduction channels and one horizontal conduction channel. The better gate-to-channel electrostatic coupling compared to a FinFET allows relaxing the scaling rules; a Triple-gate transistor can have its Fin width and height *roughly* of the dimension of the gate length. The question of the electrostatic behaviour of the corners is a potential problem because it could lead to a prior activation of corners in the structure and thus to corner leakages. However, it has been shown that this effect was insignificant for undoped Triple-gate transistors [41][42].

If the Buried OXide (BOX) is recessed due to the Fin overetch and if the gate penetrates into the BOX, the transistors are named Pi-gateFETs or ΠFETs ([43], Fig. 13.a). As well, if additionally the gate penetrates also under the silicon body, the transistors are named Omega-gateFETs or ΩFETs ([44], Fig. 13.a and Fig. 13.b). The penetration of the gate in the BOX is beneficial, as it increases the electrostatic control of the gate on the channel and therefore improves the short channel performance (Fig. 14, [43][45]).

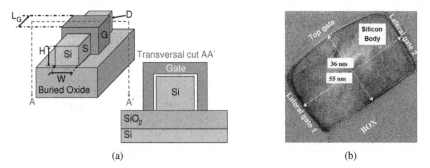

(a)   (b)

Fig. 12: Sketch of a Triple-gate transistor (a) and TEM transversal cut picture of a Triple-gate transistor (b) (from [40]).

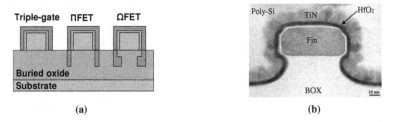

(a)   **(b)**

Fig. 13: Sketch of a TGFET, ΠFET and ΩFET transistors (a) and TEM transversal cut of an ΩFET transistor (from CEA-LETI Grenoble, [46]) (b).

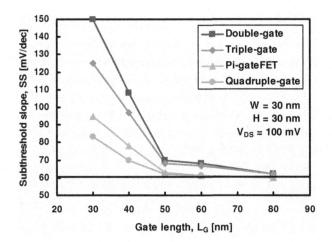

Fig. 14: Comparison of the subthreshold slope vs. gate length L_G for Double-gate, Triple-gate, ΠFET, and Quadruple-gate transistors (from [43]).

3.1.2. *Process*

As an example, the main process steps of ΩFET transistors on SOI wafer (transistors from CEA-LETI Grenoble, France, process details and performance described in [46]) are described in the Fig. 15.

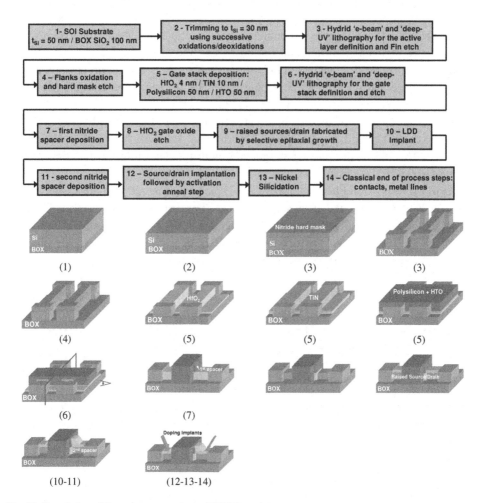

Fig. 15: Description of the main process steps of ΩFET transistors.

After the oxidation/deoxidation step in order to obtain the wished film thickness (steps 1-2), the active layer and silicon Fin are etched (usually using deep-UV or e-beam lithography, step 3). In order to decrease the defects density induced by the etch step on the lateral sides, a sacrificial oxidation/deoxidation is performed (step 4). Then, the stripping of the nitride hard mask and deposition of the gate oxide steps are made (here, 4 nm of HfO_2 deposited with Atomic Layer Deposition), and the gate material (here, titanium nitride TiN with a thickness of roughly 10 nm, using Chemical Vapor

Deposition) is deposited (step 5). The gate stack is completed by a polysilicon layer (here, 50 nm) and by an HTO (High Temperature Oxide, thickness of 50 nm). The gate stack is etched (step 6), and nitride spacers of 15 nm are deposited before the gate oxide etch (steps 7-8).

The raised source/drain regions are made using SEG (Selective Epitaxial Growth, step 9). A LDD (Lightly Doped Drain) is added in order to reduce the electric field induced by the source/substrate and drain/substrate junctions; it is done with a narrow and lightly doped region and allows reducing the SCEs (Short Channel Effects) (step 10). A second nitride spacer is deposited in order to push away the source drain junction and to control the gate leakage without degrading too much the saturation currents (step 11). The overlap capacitance between the gate and the extensions is therefore reduced, but at the cost of an increase of the effective gate length. Then, the source/drain regions are implanted (step 12), and a spike anneal is performed, followed by a nickel silicidation (step 13). The end of the process is standard, with the deposition of the contacts and metal lines (level 1) (step 14).

TEM (Transmission Electronic Microscope) pictures of the devices are shown in Fig. 16. The conformity of the gate stack is good (transversal cut, Fig. 16.a). The Ω configuration (Fig. 16.a) is visible. In the longitudinal cut (along the source/drain axis, Fig. 16.b), the BOX (Buried OXide), the silicon fin, the two nitride spacers, the silicided source/drain regions and the gate stack are visible as well.

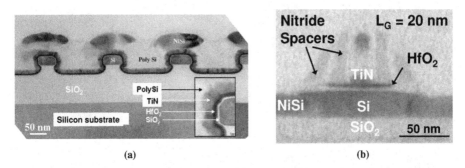

(a)   (b)

Fig. 16: TEM pictures of ΩFET transistors. (a): transversal cut of a transistor array (W = 60 nm). The inset shows the gate stack (interfacial SiO$_2$/HfO$_2$/TiN/Polysilicon). (b): longitudinal cut (along the source/drain axis) of a 20 nm gate length transistor. The gate stack, the two spacers and the silicidation are noticeable.

3.1.3. *Modeling considerations*

The modeling of TGFETs (or its variants such as ΠFET and ΩFET) structures should be 2-dimensional for long channels or 3-dimensional for short channels. As compared to double-gate transistors, the presence of a BOX brings an additional electrostatic coupling component from the back-gate.

In the section dedicated to long channels, the threshold voltage modeling of ΩFET transistors is presented; the model can be extended to the case of ΠFETs, Triple-gate FETs and planar Fully Depleted SOI (planar FDSOI) structures. The back-interface parasitic activation is highlighted, and the I_{OFF} degradation resulting from the back-interface activation. Some ways to alleviate this problem and the impact of the channel thickness are finally discussed.

In the section dedicated to short channels, an electrostatic potential expression is derived for short-channel ΠFETs and TGFETs. The solution of the 3D analytical potential is presented and the approximations made in order to obtain the analytical subthreshold current are presented and validated. The subthreshold current model is then compared with experimental measurements, in terms of subthreshold slope, threshold voltage ('Roll-off') and DIBL. The scaling of a wide range of MuGFET devices is then presented, and a practical pseudo-compact equation for the subthreshold slope is proposed.

3.2. *Long channels*

3.2.1. *Interface coupling model in ΩFET transistors*

Assumptions

Due to the impossibility to dope uniformly the silicon Fin leading to unacceptable threshold voltage variability, the channel is left undoped in vertical MuGFETs transistors (implying the use of a metal gate). As a result, the interface coupling between front- and back- gates can be modelled using the 2D Poisson's equation. Below threshold, the minority carrier concentration can be ignored; it is considered that the subthreshold approximation is valid up to threshold. Therefore, for undoped channels and subthreshold operation, the 2D Poisson's equation describing the electrostatics can be approximated by the Laplace's equation:

$$\frac{\partial^2 \psi(x, y)}{\partial x^2} + \frac{\partial^2 \psi(x, y)}{\partial y^2} = 0 \tag{59}$$

with the boundary conditions determined by the surface potential and the axis defined in Fig. 17.

Additionally, the surface potential along the top and lateral gates will be considered constant (φ_{S1}, see Fig. 17) all along the Triple-gate body/gate oxide [47]. This approximation is not perfectly accurate since the surface potential is also influenced by the back-gate (φ_{S1} exhibits variations from the bottom to the top of the lateral gates). However, as the threshold occurs when one point in the channel reaches the threshold value (φ_{ST}, which is a function of the channel doping, but is independent of the width and height of the Fully Depleted channel [47]), a constant φ_{S1} is an acceptable approximation to model the threshold voltage in undoped devices. This approximation is also coherent with neglecting corner effects since we will assume that the surface potential at the corners is equal to its value at x = W/2 (Fig. 7, [38]).

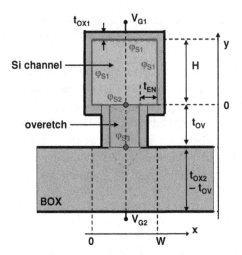

Fig. 17: Sketch of the transversal cross-section of an ΩFET with the notations used in this work. A TGFET corresponds to $t_{OV} = 0$ nm and $t_{EN} = 0$ nm, and a ΠFET to $t_{OV} \neq 0$ nm and $t_{EN} = 0$ nm.

Considering the boundary conditions of the potential at the body/overetched region and overetched region/BOX interface, the potential will be supposed parabolic (with a value of φ_{S1} at the lateral interfaces or all along the penetration of the gate under the body, and a value of respectively φ_{S2} ($x = W/2$, $y = 0$) and φ_{S3} ($x = W/2$, $y = -t_{OV}$) at mid-channel (Fig. 17)). The parabolic approximation for the back-interface boundary condition is very appropriate for narrow Fins, and acceptable for wide Fins (where a constant back-interface potential would be more exact since the influence of the lateral gates is negligible); however, for the same reasons given for the front-surface potential, it could be approximated that the first point in the back-channel to reach threshold gives a good idea of the global threshold voltage of the whole channel. Considering channel widths W and heights H > 10 nm, the quantum effects will also be neglected [48].

Potential:

First, we consider that the potential is equal to 0 on the top, left and right interfaces. Applying the separation of variables [49], the potential is given by:

$$\psi(x,y) = X(x)Y(y) \tag{60}$$

Therefore, the Laplace's equation Eq.(59) yields:

$$\frac{\partial^2 X/\partial x^2}{X} = \lambda = \frac{-\partial^2 Y/\partial y^2}{Y} \tag{61}$$

with λ a real number.

The potential can therefore be written as:

$$X(x) = C_1 e^{i\sqrt{\lambda}x} + C_2 e^{-i\sqrt{\lambda}x} \tag{62}$$

$$Y(y) = C_3 e^{\sqrt{\lambda}y} + C_4 e^{-\sqrt{\lambda}y} \qquad (63)$$

with C_1, C_2, C_3, and C_4 parameters to be calculated using the boundary conditions.

Furthermore, $X(0)=0$, and therefore $C_1=-C_2$.
Thus, $X(x)$ can be written as:

$$X(x) = A\sin(\lambda x) \qquad (64)$$

with A a constant.
Additionally, $X(W)=0$, implying that:

$$\lambda_n = (\frac{n\pi}{W})^2 \qquad (65)$$

with n a natural number.

Therefore, applying the linear superposition the potential solution is given by:

$$\psi(x, y) = \sum_{n=1}^{+\infty} Y(y)\sin(\frac{n\pi x}{W}) \qquad (66)$$

Furthermore, $Y(H)=0$ and therefore:

$$\psi(x, y) = \sum_{n=1}^{+\infty} L_n \text{sh}(\frac{n\pi(H-y)}{W})\sin(\frac{n\pi x}{W}) \qquad (67)$$

with L_n a constant.

The constants L_n are then determined by the final boundary condition:

$$\psi(0, y) = \varphi_0(y) = \begin{cases} 0 \text{ if } 0 < y < t_{EN} \text{ and } W - t_{EN} < y < W \\ \text{parabola if } t_{EN} < y < W - t_{EN} \end{cases} \qquad (68)$$

with the parabola taking the value of 0 at $y = t_{EN}$ and $y = W - t_{EN}$, and $\varphi_{S,MID}$ the extreme value (at $x = W/2$) of the potential at the bottom interface.

Eq.(67) at $y=0$ is a Fourier sine series for the function $\varphi_0(y)$ defined for $0<y<W$ with Fourier coefficients $L_n \text{sh}(n\pi H/W)$ [49]:

$$L_n \text{sh}(\frac{n\pi H}{W}) = \frac{2}{W}\int_0^W \varphi_0(y)\sin(\frac{n\pi y}{W})dy \qquad (69)$$

Therefore, it finally yields:

$$L_n = \varphi_{S,MID}\frac{F_n}{\text{sh}(\frac{n\pi H}{W})} \qquad (70)$$

with F_n defined equal to:

$$F_n(W, H, e_{EN}) =$$

$$W \frac{\left[\begin{array}{l} (2t_{EN} n\pi \sin(n\pi) - n\pi W \sin(n\pi) - 2W \cos(n\pi) + 2W) \cos(\dfrac{n\pi t_{EN}}{W}) + \\ (-2t_{EN} n\pi \cos(n\pi) + n\pi W \cos(n\pi) + 2t_{EN} n\pi - n\pi W) \sin(\dfrac{n\pi t_{EN}}{W}) \end{array} \right]}{(\dfrac{n\pi}{2})^3 (W - 2t_{EN})^2} \tag{71}$$

Considering now that the potential is constant and equal to φ_{S1} at the top and lateral interfaces, the potential in the silicon body will be simply given changing the constant value of the potential:

$$\psi_{Si}(x, y) = \varphi_{S1} + (\varphi_{S2} - \varphi_{S1})$$

$$\sum_{n=1}^{+\infty} F_n \frac{sh(\dfrac{n\pi(H - y)}{W})}{sh(\dfrac{n\pi H}{W})} \sin(\dfrac{n\pi x}{W}) \tag{72}$$

with F_n defined in Eq.(71).

As in the channel, similar calculations with two parabolas as boundary conditions at the top and bottom interfaces yield for the overetched region:

$$\psi_{OV}(x, y) = \varphi_{S1} +$$

$$\left[P_n \sin(\dfrac{n\pi x}{W_2}) \sum_{n=1}^{+\infty} \frac{(\varphi_{S2} - \varphi_{S1}) sh(\dfrac{n\pi y}{W_2}) + (\varphi_{S3} - \varphi_{S1}) sh(\dfrac{n\pi(t_{OV} - y)}{W_2})}{sh(\dfrac{n\pi t_{ov}}{W_2})} \right] \tag{73}$$

with $W_2 = W - 2t_{EN}$ the width of the overetched region and:

$$P_n = \frac{2(1 - \cos(n\pi)) - n\pi \sin(n\pi)}{(\dfrac{n\pi}{2})^3} \tag{74}$$

It should be noted that the difference between F_n and P_n defined in Eq.(71) and Eq.(74) lies in the difference of boundary conditions for the two regions: taking the "Ω" shape in the channel (Fig. 17), and parabolic in the overetched region.

The links between the surface potentials and the front- and back-gate biases is made considering the Gauss's law at the interfaces.

At the gate oxide/silicon interface, the continuity of potentials is given by:

$$V_{G1} = V_{FB1} + \varphi_{S1} + \Delta V \tag{75}$$

with ΔV being the voltage drop in the gate oxide.

Applying the Gauss's law in the gate oxide yields:

$$V_{G1} = V_{FB1} + \varphi_{S1} + \frac{\varepsilon_{Si}}{C_{OX1}} E(\frac{W}{2}, H) \tag{76}$$

with $C_{OX1} = \varepsilon_{OX1}/e_{OX1}$ the gate oxide capacitance and $E(W/2, H)$ the electric field at the gate oxide/silicon interface and at mid-channel.
Therefore, using Eq.(72) in Eq.(76) yields:

$$V_{G1} = V_{FB1} + \varphi_{S1}(1 + B) - \varphi_{S2}B \tag{77}$$

with:

$$B(W, H) = \frac{C_{LAT}}{C_{OX1}} \sum_{n=1}^{+\infty} \frac{n\pi F_n}{\text{sh}(\frac{n\pi H}{W})} \sin(\frac{n\pi}{2}) \tag{78}$$

with $C_{LAT} = \varepsilon_{Si}/W$ a lateral capacitance in the body.

For the back-interface, a similar derivation from Eq.(73) yields:

$$V_{G2} = V_{FB2} + \varphi_{S3} + C(\varphi_{S2} - \varphi_{S1}) + D(\varphi_{S1} - \varphi_{S3}) \tag{79}$$

with:

$$C(W, t_{ov}) = \frac{C_{LAT}^{BOX}}{C'_{OX2}} \sum_{n=1}^{+\infty} \frac{n\pi P_n}{\text{sh}(\frac{n\pi t_{ov}}{W_2})} \sin(\frac{n\pi}{2}) \tag{80}$$

and:

$$D(W, t_{ov}) = \frac{C_{LAT}^{BOX}}{C'_{OX2}} \sum_{n=1}^{+\infty} \frac{n\pi P_n}{\text{th}(\frac{n\pi t_{ov}}{W_2})} \sin(\frac{n\pi}{2}) \tag{81}$$

with $W_2 = W - 2t_{EN}$ the width of the overetched region, $C_{LAT}^{BOX} = \varepsilon_{OX2}/W_2$ a lateral capacitance in the BOX, and $C'_{OX2} = \varepsilon_{OX2}/(t_{OX2} - t_{ov})$ the BOX capacitance (minus the overetched zone).

The last relationship between φ_{S1}, φ_{S2}, and φ_{S3} is obtained from Eq.(72) and Eq.(73) by considering the Gauss's law at the silicon/overetched interface:

$$E.(\varphi_{S2} - \varphi_{S1}) = G.(\varphi_{S1} - \varphi_{S2}) + F.(\varphi_{S3} - \varphi_{S1}) \tag{82}$$

with:

$$E(W,H) = C_{LAT} \sum_{n=1}^{+\infty} \frac{n\pi F_n}{th(\frac{n\pi H}{W})} \sin(\frac{n\pi}{2}) \tag{83}$$

and:

$$F(W,t_{OV}) = C_{LAT}^{BOX} \sum_{n=1}^{+\infty} \frac{n\pi P_n}{sh(\frac{n\pi t_{OV}}{W_2})} \sin(\frac{n\pi}{2}) \tag{84}$$

and:

$$G(W,t_{OV}) = C_{LAT}^{BOX} \sum_{n=1}^{+\infty} \frac{n\pi P_n}{th(\frac{n\pi t_{OV}}{W_2})} \sin(\frac{n\pi}{2}) \tag{85}$$

Finally, using Eq.(79) and Eq.(82) yields:

$$V_{G2} = V_{FB2} + (C - D - (\frac{1+D}{F})(E + G - F))\varphi_{S1}$$
$$+ ((\frac{1+D}{F})(E + G) - C)\varphi_{S2} \tag{86}$$

with C, D, E, F, and G defined in Eq.(80) to Eq.(85).

Eq.(77) and Eq.(86) are the two equations allowing to derive the threshold voltage models in order to obtain the threshold voltage of the front- (V_{TH1}) and back-gates (V_{TH2}).

Front-gate threshold voltage V_{TH1}

Splitting the three regimes of the back-interface (i.e. accumulation, depletion and inversion) is the traditional way to obtain a general threshold voltage model in Fully Depleted structures [47,50]. The calculation of the back-interface accumulation $V_{G2,ACC2}$ and inversion $V_{G2,INV2}$ voltages is made by considering the limit values of the surface potentials φ_{S1} and φ_{S2} (Eq.(77) and Eq.(86)). Next, the front-gate threshold voltage V_{TH1} is derived accordingly with the regime of the back-gate (Eq.(87) and Eq.(88)). The threshold voltage is constant when the back-interface is accumulated $V_{TH1,ACC2}$ or inverted $V_{TH1,INV2}$, and decreases linearly with V_{G2} when the back-interface is depleted (Fig. 18).

Therefore, considering the three regimes of the back-gate separately:

a) Back-gate accumulated:

In this case, $\varphi_{S1}=\varphi_{ST}$ and $\varphi_{S2}=0$. Therefore, Eq.(86) yields:

$$V_{G2,ACC2} = V_{FB2} + (C - D - (\frac{1+D}{F})(E + G - F))\varphi_{ST} \tag{87}$$

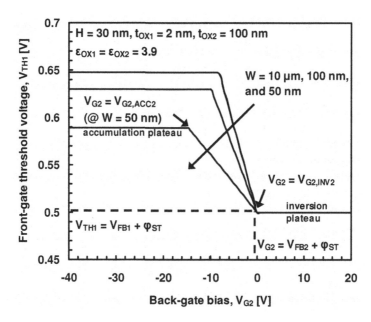

Fig. 18: Model of the *front-gate* threshold voltage V_{TH1} vs. back-gate bias V_{G2} for TGFET transistors for various fin widths W. Fin height H =30 nm, gate oxide thickness t_{OX1}=2 nm, BOX thickness t_{OX2}=100 nm.

b) Back-gate inverted:
With an inverted back-interface, $\varphi_{S1} = \varphi_{S2} = \varphi_{ST}$.
Eq.(86) yields:

$$V_{G2,INV2} = V_{FB2} + \varphi_{ST} \tag{88}$$

c) Back-gate depleted:
The back-gate is depleted when $V_{G2,ACC2} < V_{G2} < V_{G2,INV2}$.

The front-gate threshold voltage V_{TH1} can now be obtained:
a) Back-gate accumulated ($V_{G2,ACC2} < V_{G2}$):
Since $\varphi_{S1} = \varphi_{ST}$ and $\varphi_{S2} = 0$, Eq.(77) yields:

$$V_{TH1,ACC2} = V_{FB1} + (1+B)\varphi_{ST} \tag{89}$$

b) Back-gate inverted ($V_{G2} > V_{G2,INV2}$):
Since $\varphi_{S1} = \varphi_{S2} = \varphi_{ST}$, Eq.(77) yields:

$$V_{TH1,INV2} = V_{FB1} + \varphi_{ST} \tag{90}$$

c) back-gate depleted ($V_{G2,ACC2} < V_{G2} < V_{G2,INV2}$):
The threshold voltage is a function of the back-gate bias V_{G2} since the back-gate is depleted and Eq.(77) to Eq.(86) yield:

$$V_{TH1,DEP2}(W, H, V_{G2}) = V_{FB1} -$$

$$(\frac{B}{1 + \frac{1+D}{F}(E+G-F) - C + D}).(V_{G2} - V_{FB2})$$
(91)

$$+ (1 + \frac{B}{1 + \frac{1+D}{F}(E+G-F) - C + D}).\varphi_{ST}$$

Back-gate threshold voltage V_{TH2}

Similar considerations as in the front-gate are applied to the back-gate in order to yield the back-gate threshold voltage:

a) Front-gate accumulated:

We have $\varphi_{S2}=\varphi_{ST}$ and $\varphi_{S1}=0$. Therefore, Eq.(77) yields:

$$V_{G1,ACC1} = V_{FB1} - B\varphi_{ST}$$
(92)

b) Front-gate inverted:

We now have $\varphi_{S1}= \varphi_{S2}=\varphi_{ST}$. Eq.(77) yields:

$$V_{G1,INV1} = V_{FB1} + \varphi_{ST}$$
(93)

c) Front-gate depleted:

The front-gate is depleted when $V_{G1,ACC1}<V_{G1}<V_{G1,INV1}$.

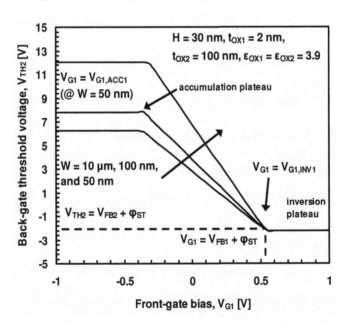

Fig. 19: Model of the *back-gate* threshold voltage V_{TH2} vs. back-gate bias V_{G1} for TGFET transistors for various fin widths W. Fin height H =30 nm, gate oxide thickness t_{OX1}=2 nm, BOX thickness t_{OX2}=100 nm.

The back-gate threshold voltage V_{TH2} can now be obtained:

a) Front-gate accumulated ($V_{G1,ACC1}<V_{G1}$):

Since $\varphi_{S2}=\varphi_{ST}$ and $\varphi_{S1}=0$, Eq.(86) yields:

$$V_{TH2,ACC1} = V_{FB2} + ((1+D)(\frac{E+G}{F}) - C)\varphi_{ST} \tag{94}$$

b) Front-gate inverted ($V_{G1}>V_{G1,INV1}$):

Since $\varphi_{S1}= \varphi_{S2}=\varphi_{ST}$, Eq.(86) yields:

$$V_{TH2,INV1} = V_{FB2} + \varphi_{ST} \tag{95}$$

c) Front-gate depleted ($V_{G1,ACC1}<V_{G1}<V_{G1,INV1}$):

The threshold voltage is a function of the front-gate bias V_{G1} since the front-gate is depleted and Eq.(77) to Eq.(86) yield:

$$V_{TH2,DEP1}(W,H,V_{G1}) = V_{FB2} -$$

$$(\frac{(1+D)(\frac{E+G}{F}) - C - 1}{1+B}).(V_{G1} - V_{FB1})$$

$$+ (1 + \frac{(1+D)(\frac{E+G}{F}) - C - 1}{1+B}).\varphi_{ST} \tag{96}$$

Complete interface model $V_{TH,TOT}$

The expressions derived in the previous sections for the threshold voltage of the front- and back-gate model the influence of one interface upon the other (interface coupling). However, in a Fully Depleted structure, it is a well known fact [51] that in the back-interface inversion regime the first channel to invert is the back-channel, followed by the front-channel.

Therefore, both channels (front- and back- channels) and their respective threshold voltages should be merged in order to build a complete threshold voltage model of these structures. Plotting along a "V_{G1}-V_{G2}" axis, four zones can be defined: no channel inverted, one of the two channels inverted, and both channels inverted (Fig. 20). In an asymmetrical structure like an ΩFET, the threshold voltage of a channel is also a function of the bias applied on the other gate. If the back-gate bias V_{G2} is kept constant while varying the front-gate bias V_{G1}, the threshold 'as seen during measurements' is a combination of the activation of front- and back-channel. For $V_{G2}<V_{G2,INV2}$, only one threshold voltage is noticeable, and is related to the inversion of the front-channel. Because of the coupling between the two channels, for $V_{G2}<V_{G2,INV2}$ the first channel to invert while ramping the front-gate bias is the back-channel. Using extraction methods like the constant current method or the maximum of transconductance method, only this threshold can be extracted. However, using other methods like the double derivative of the drain current (Fig. 21), the activation of the front-interface (horizontal dashed line, Fig. 20) becomes visible (cf. Fig. 21 and transconductance humps in Fig. 26).

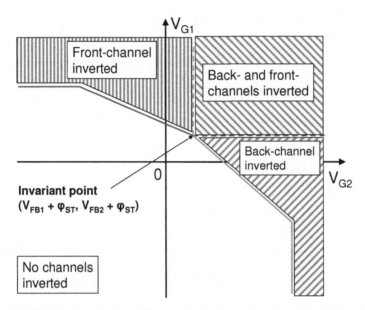

Fig. 20: Zones highlighting the activation of front- and back- channels as a function of front- and back- gate biases. The position and shapes of the zones vary with the flat-band voltages V_{FB1} and V_{FB2}, width W, height H, and gate oxide / BOX thicknesses t_{OX1} and t_{OX2}.

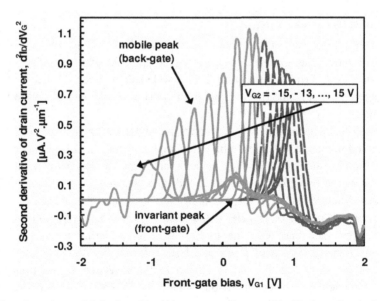

Fig. 21: Experimental second derivative of the drain current vs. front-gate bias V_{G1} for various back-gate biases V_{G2} ($V_{G2} < - 3V$ (dashed lines); $V_{G2} > -3$ V (solid lines)). $L_G = 10$ μm, W = 10 μm, H = 26 nm, and $V_{DS} = 10$ mV.

The threshold voltage of the structure is given by Eq.(87) to Eq.(91) for front-interface and by Eq.(92) to Eq.(96) for the back-interface.

Therefore, it yields:
For $V_{G2,INV2} < V_{G2} < V_{G2,INV2} + V_{G2,ACC2}$:

$$V_{TH2,DEP1}(W,H,V_{G2}) = V_{G2,INV2} -$$
$$(\frac{1+B}{(1+D)(\frac{E+G}{F})-C-1}).(V_{G2} - V_{FB2}) \qquad (97)$$

with V_{TH2} leaning toward $+\infty$ (resp. $-\infty$) for V_{G2} leaning toward $V_{G2,INV2}$ (resp. $V_{G2,ACC2}$).

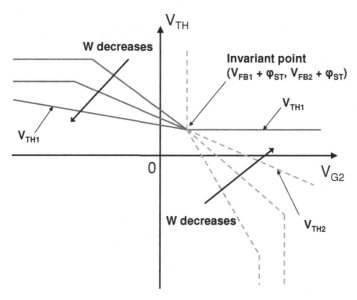

Fig. 22: Total threshold model V_{TH} (V_{TH1}: front-gate, solid lines; V_{TH2}: back-gate, dashed lines) vs. back-gate bias V_{G2}. The evolution while decreasing the gate width W is also shown.

The evolution of the total threshold voltage V_{TH} (V_{TH1} when the back-interface is accumulated or depleted, two threshold voltages V_{TH1} and V_{TH2} when the back-interface is inverted) is given in Fig. 12. In a transistor featuring a channel width W much larger than the channel height H (*i.e.*, behaving as a planar Fully Depleted SOI (FDSOI) MOSFET, Fig. 8), the threshold voltage is solely a function of the Fin height [50]. Compared to the FDSOI configuration (W =10 μm), a TGFET transistor (Fig. 18) has, when the fin width W is decreased, the back-gate accumulation bias $V_{G2,ACC2}$ shifted toward more negative biases and a smaller front-channel threshold voltage $V_{TH1,ACC2}$. In other words, the control from the lateral gates on φ_{S2} leads to a more negative back-gate bias required to create the accumulation of the back-channel. As the back-gate has less influence than for wider transistors, the electrostatic control of the front-gate on the channel is enhanced and the front-channel threshold voltage V_{TH1} is smaller than for wide transistors. Subsequently, the front-gate to back-gate coupling coefficient (defined as the slope of the $V_{TH1}(V_{G2})$ curve when the back-channel is depleted) decreases if the Fin width is reduced: the lateral electrostatic coupling screens the vertical coupling induced by the back-gate. Similarly, the back-interface activation is effectively screened by the lateral coupling and occurs therefore for a voltage closer to the front-interface threshold. The threshold voltage model for ΩFETs can be extended to other structures, like ΠFETs, TGFETs, and planar FDSOI structures (Tab. 1).

Table 1. Variations of the core structure

Structure	Features
ΩFET (core structure)	$t_{OV} \neq 0$, $t_{EN} \neq 0$
ΠFET	$t_{OV} \neq 0$, $t_{EN} \approx 0$
TGFET	$t_{OV} \approx 0$, $t_{EN} \approx 0$
Planar FDSOI	$t_{OV} \approx 0$, $t_{EN} \approx 0$, W>>H

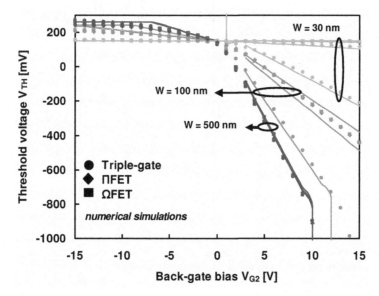

Fig. 23: Numerical simulations of the extracted threshold voltage (symbols, constant current method) versus substrate voltage V_{G2} for Tri-gate (circles), ΠFETs (triangles) and ΩFETs (squares), and comparison with the analytical model (solid lines). Gate widths W are 30, 100 and 500 nm. Drain voltage V_{DS} is 50 mV and gate length L_G is 10 μm. $t_{OX1} = 2$ nm, $t_{OX2} = 100$ nm, H = 26 nm, $t_{OV} = 30$ nm, $t_{EN} = 5$ nm.

In Fig. 23, numerical simulations (using Silvaco ATLAS [52]) of threshold voltage versus back-interface voltage curves are plotted for various gate widths and transistor configurations (Triple-gate, ΠFET and ΩFET geometries). For wide devices (W = 500 nm) where the influence of the lateral gates is negligible, the characteristics of all three devices are identical, given the fact that the interface coupling is in this case only vertical (between front- and back-gates).

As the gate width is reduced, the lateral coupling tends to screen the threshold voltage from the back-gate voltage as observed in Fig. 13. The threshold voltage of the Tri-gateFET is far more sensitive to back-gate bias than for both ΠFETs and ΩFETs. This difference is clearly visible for very narrow channels (W = 30 nm), and is due to the better isolation of ΠFETs/ΩFETs to the back-gate bias thanks to the penetration of the front-gate in the BOX. Thanks to the overetch of the lateral gates, the ΠFET and even better the ΩFET structures act like 'quasi gate-all-around' transistors. The general agreement between the model and the numerical simulations is excellent.

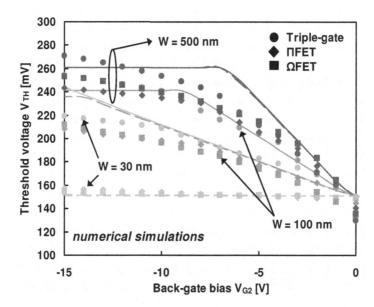

Fig. 24: Numerical simulations of the extracted threshold voltage (symbols, constant current method) versus substrate voltage V_{G2} for Tri-gate (circles), ΠFETs (triangles) and ΩFETs (squares), and comparison with the analytical model (lines). Gate widths W are 30, 100 and 500 nm. Drain voltage V_{DS} is 50 mV and gate length L_G is 10 μm. t_{OX1} = 2 nm, t_{OX2} = 100 nm, H = 26 nm, t_{OV} = 30 nm, t_{EN} = 5 nm.

Fig. 24 is a zoom of Fig. 23, in the zones of 'back-interface depletion' and 'back-interface accumulation'. The difference between TGFETs and ΠFETs/ΩFETs lies in the fact that the screening of the back-gate by the coupling from the lateral gates increases in ΠFET/ΩFETs structures due to the overetched region. Subsequently, the accumulation is reached for even more negative back-gate biases. This leads to a smaller coupling coefficient, or in other words to a larger insensitivity to the back-gate bias for ΠFETs/ΩFETs structures. As a matter of fact, the electrostatic control of the front-gate on the channel is enhanced in ΠFET/ΩFET transistors compared to TGFET transistors. The general agreement between the model and the numerical simulations is good.

3.2.2. Effect on the interface conductions

The model has been compared with experimental ΩFETs transistors (Fig. 15, process details and performance described in [46]). The agreement between the model and the experimentally extracted threshold voltage is very good. The invariant point predicted by the analytical model is clearly noticeable in the experimental measurements (Fig. 25) and occurs when $V_{G2} = V_{FB2} + \varphi_{ST}$. At this point corresponding to the brink of back-interface inversion, the surface potential at the back-gate interface φ_{S2} attains a value compensating the vertical electric field, leading to a flat potential in the channel at the front-gate threshold voltage.

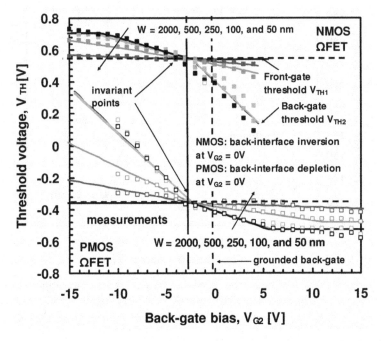

Fig. 25: Experimental threshold voltage (constant current method, squares) as a function of back-gate bias V_{G2} and fin width W (from W = 2 μm down to W = 50 nm) for NMOS and PMOS and comparison with the analytical model (solid lines). L_G = 10 μm, H = 26 nm, t_{OX2} = 100 nm, and V_{DS} = 50 mV.

For these transistors, the flat-band voltage of the back-channel V_{FB2} is extracted with the invariant point (Fig. 25) and is around -3 V. Therefore, any measurement with V_{G2} grounded corresponds to the back-interface inversion for NMOS, and to the back-interface depletion for PMOS. The threshold voltage V_{TH} at $V_{G2} = 0$ V increases when the fin width W is reduced for NMOS (Fig. 25). This seemingly illogical result (compared to electrostatics considerations) lies in the fact that the transistor operates in back-channel inversion and that the increase of threshold voltage when reducing the channel width is due to a progressive alignment of the front- and back-channel threshold voltages, all four channels being activated at V_{G2} grounded.

Plotting the transconductance of the experimental devices (Fig. 26), two peaks are noticeable for wide NMOS transistors. The first peak is related to the inversion of the back-interface and therefore moves when changing the back-gate bias V_{G2}, while the second peak is related to the inversion of the front-interface and is independent of the back-gate bias (in coherence with the compact model). For narrow NMOS transistors, the two threshold voltages merge and only one threshold is noticeable. Using the same base wafers, as the NMOS devices operates in back-interface inversion the PMOS devices should be in back-interface depletion at V_{G2} grounded (Fig. 25). Therefore, only one peak is visible for PMOS transistors. As well, the increase of threshold voltage when the gate width is reduced for both NMOS and PMOS at $V_{G2} = 0$ V (Fig. 26) is related to the fact that the back-interface is inverted for NMOS and depleted for PMOS at $V_{G2} = 0$ V (Fig. 25).

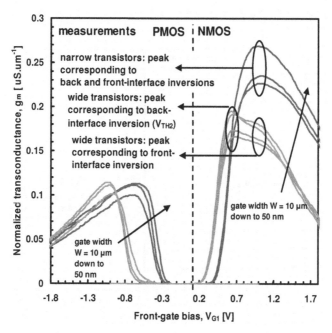

Fig. 26: Experimental normalized transconductance g_m vs. front-gate bias V_{G1} and gate width W for NMOS/PMOS. $V_{G2} = 0$ V, $L_G = 10$ μm, $e_{OX2} = 100$ nm, H = 26 nm, and $V_{DS} = 50$ mV.

The consequence is that the width normalization is to be considered carefully, since it has to be done using the effective width $W_{EFF}=2(H+W)$ for back-channel inversion regime and using $W_{EFF}=2H+W$ for back-channel depletion/accumulation regimes, and not always $W_{EFF}=2H+W$ as usually assumed. These points can affect considerably the extracted mobility for wide transistors (FDSOI transistors configuration). Using the $V_{TH1}(V_{G2})$ curves for various fin widths, the position of the invariant point allows determining directly the regime of the back-interface (depleted or inverted) when the back-interface is grounded (i.e. for $V_{G2} = 0$ V). Additionally, compact models of Planar

Single-gate FDSOI transistors must take into account the back-interface conduction (even tough the Planar Single-gate FDSOI transistors are called 'single-gate' transistors) to correctly reproduce the behaviour of the transistor, especially for the threshold voltage (like the approach presented in this chapter, section 2. and in [53]).

Case of FDSOI devices

- Minimization of gate leakage:

In ΩFET devices (and therefore in ΠFETs and TGFETs) with a narrow width (W < 30 nm), the effect of the back-gate is practically deactivated. It means that the back-interface conduction exists, but with a threshold voltage extremely close to the threshold voltage of the front-gate, and a magnitude much smaller due to the much thicker thickness of the Buried Oxide. The back-interface inversion effect is maximal for Planar FDSOI structures, where the effect of the back-gate bias is unscreened by the lateral gates (like in TGFETs/ΠFETs/ΩFETs).

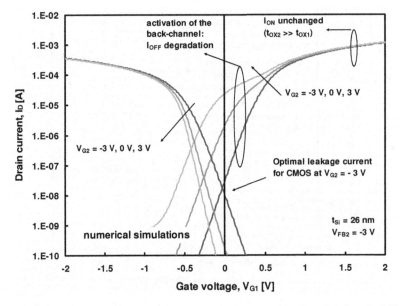

Fig. 27: Simulated Drain current I_D vs. front-gate bias V_{G1} for back-gate bias V_{G2} of -3, 0, and +3 V, and for PMOS and NMOS. Silicon thickness H = 26 nm, V_{FB2} = - 3 V.

Using the same parameters as for the experimental devices (H = 26 nm, V_{FB2} = -3 V), NMOS and PMOS are simulated for short gate lengths (L_G = 40 nm, Fig. 27). For NMOS, the onset of the back-channel is visible for V_{G2} = 0 and +3 V. As the threshold voltage of the front-gate does not change, it leads to a dramatic increase of the current in the OFF state (I_{OFF}). For PMOS devices, no increase (even, a reverse trend) is noticed as the back-gate is depleted. Operating near the invariant point, it is possible to optimize the performance of both NMOS and PMOS simultaneously.

Considering short channels, the back-interface conduction might mean a short cut between source and drain and therefore a considerable degradation of the short channel performance. In Fig. 28, it can be seen that the back-channel inversion induces a degradation of subthreshold slope and DIBL, comparing the operation at $V_{G2} = 0$ V and at -3 V (invariant point). The simulation results for the subthreshold slope matches the experimental data (diamonds, Fig. 28). However, the degradation of subthreshold slope/DIBL is not as important as the I_{OFF} degradation, which will be an effect present in both short and long channels.

One solution to suppress this effect would be to force the back-interface into depletion for both NMOS and PMOS (i.e. close to the invariant point, Fig. 25) by applying a negative bias at the back-gate, but it would require a bias of several volts. The most pragmatic solution in order to mitigate the influence of the back-gate is however to use a thick BOX (see next section).

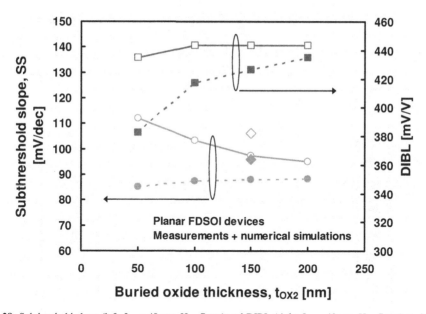

Fig. 28: Subthreshold slope (left, $L_G = 40$ nm, $H = 7$ nm) and DIBL (right, $L_G = 40$ nm, $H = 7$ nm) vs. buried oxide thickness t_{OX2} for $V_{G2} = -3$ V (closed circles) and 0 V (open circles). For the subthreshold slope are represented the simulated (squares) and experimental (diamonds). $V_{FB2} = -3$ V, $t_{OX1} = 1.65$ nm, $V_{DS} = 50$ mV.

Influence of channel thickness

Changing the channel thickness in UTB SOI planar transistors modifies the interface coupling. Comparing the model for long channels with experimental measurements featuring different channel ($t_{Si} = 26, 13, 7$ nm), gate oxide (t_{OX1} (EOT) = 1.95, and 1.65 nm), and BOX thicknesses ($t_{OX2} = 145$, and 100 nm), a correct agreement with the model is found (Fig. 29).

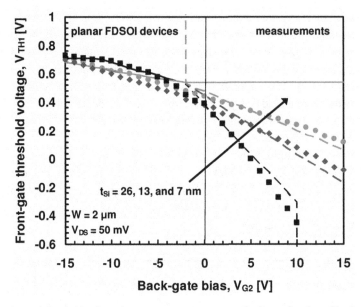

Fig. 29: Experimental threshold voltage (squares, constant current method) as a function of back-gate bias V_{G2} and channel thickness t_{Si} (26, 13, and 7 nm) and comparison with the analytical model (lines). $L_G = 10$ μm, $W = 10$ μm, $t_{OX1} = 1.95/1.65$ nm, $t_{OX2} = 100/145$ nm, and $V_{DS} = 50$ mV.

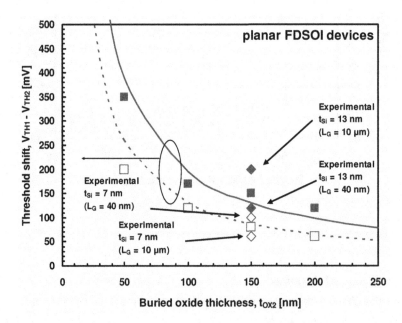

Fig. 30: Difference between front- and back-gate ($V_{TH1} - V_{TH2}$) threshold voltage vs. buried oxide thickness t_{OX2} obtained with the model (lines), simulations (squares) and experimental data (diamonds), and for $t_{Si} = 13$ nm (closed symbols) and $t_{Si} = 7$ nm (open symbols).

As the NMOS devices used in this work operate at V_{G2} grounded in the 'back-interface inversion' regime, two thresholds (front- and back- interfaces) coexist in the structures. In other words, the back-interface conduction creates a parasitic channel degrading the subthreshold slope and the I_{OFF} (the degradation is a function of the channel length and of V_{FB2}, but is typically in the order of magnitude of 1 decade per 100 mV threshold voltage $(V_{TH1}-V_{TH2})$ difference), while not improving significantly the I_{ON} (because $t_{OX2} \gg t_{OX1}$). Using the model, a formula of the threshold voltage difference between front- and back- gate is derived, and compared with simulation/experimental data obtained for a given buried oxide thickness with a good agreement (Fig. 30). Additionally, It is shown that for aggressive dimensions (planar FDSOI structures, H = 7 and 13 nm, Fig. 30), the parasitic activation of the back-interface is still noticeable. As well, the effect exists for short channels, where it sums up with classical short channel effects (open diamonds, Fig. 30).

3.3. *Short channels TGFETs*

In this section, the 3D analytical modeling of short-channels ΠFETs (and TGFETs transistors) is investigated. Based on the solution of the 3D Laplace's equation, the interface coupling in the structure is described and the potential calculated. Using the 'most leaky path' approach, the potential is then integrated and expressed as a simplified formula for the subthreshold current of ΠFET transistors. The short-channel characteristics (subthreshold current, subthreshold slope, Roll-off and DIBL) are calculated and compared to experimental data.

3.3.1. *Derivation of the potential*

Similarly as what for long channels, the Poisson's equation can be approximated by the Laplace's equation under threshold:

$$\frac{\partial^2 \varphi(x,y,z)}{\partial x^2} + \frac{\partial^2 \varphi(x,y,z)}{\partial y^2} + \frac{\partial^2 \varphi(x,y,z)}{\partial z^2} \approx 0 \tag{98}$$

with $\varphi(x,y,z)$ the electrostatic potential in the channel.

The following assumptions are being made:
- The channel doping is 10^{15} cm^{-3}, and the source and drain junctions are assumed to be abrupt.
- The influence of the field penetration from the drain and the overlapped regions of the gate through the BOX (the so-called 'DIVSB' effect [54]) is neglected.
- Considering width W and height H of the channel above 10 nm, the quantum effects are neglected; numerical simulations have clearly shown that they do not induce significant variations of threshold voltage and subthreshold slope down to W = H = 10 nm [48].

- The corner effects are also neglected, considering the use of undoped channels (demonstrated in [38][41] for TGFETs, and in [55] for ΠFETs).
- As far as the boundary conditions are concerned, the potential profile in the lateral direction is assumed parabolic both at the interfaces between channel and overetch region, and between overetch region and BOX (Fig. 31).

The equation that has to be solved is therefore a 3D Laplace's equation for the device configuration shown in Fig. 31. In order to calculate the potential in the structure, the influences of the six apparent terminals corresponding to the external boundary conditions (namely source, drain, the three sides of the front-gate, and the back-gate) are considered separately ('superposition theorem'). For each terminal, the potential is developed in Fourier series (a similar approach as in [57]) while setting the other terminals to zero. This approach has the advantage of creating symmetries and simplifying the calculation of the Fourier series coefficients (the mathematical details of the calculation can be found in [49]). The Fourier series coefficients are only functions of the considered boundary conditions (Dirichlet boundary condition with a constant or parabolic value, and Neumann boundary condition) and are given in the Appendix.

Finally, the Gauss's theorem is applied at the back-interface in order to take into account the effects of the back-gate and the Π-shape of the transistor. The obtained potential formula is given in Eq.(99) to Eq.(104).

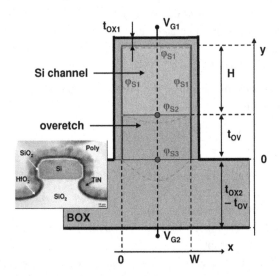

Fig. 31: Sketch of the transversal cross-section of a ΠFET with the notations used in this work. A TGFET corresponds to $t_{ov} = 0$ nm, and a ΠFET to $t_{ov} \neq 0$. Inset shows a TEM image of the devices [20]. The gate encroachment under the Si film is small enough so the ΩFET can be considered as a ΠFET.

The comparison of the analytical formula with numerical simulations (using COMSOL Femlab [56] to solve the Laplace's equation) shows an excellent agreement (Fig. 32 and Fig. 33). It can be seen that the parabolic approximation in the transversal direction yields very good results (Fig. 32) and that the set of Equations (Eq.(99) to Eq.(104)) is able to correctly reproduce the numerical simulations. Along the longitudinal axis (Source/Drain axis), the potential is very strongly dependent on the source and drain biases as well as on the lateral gates. Their influence extends in the channel as well as in the overetched region. The general agreement between model and simulation is good (Fig. 33). The small discrepancy model/simulation close to the boundaries (lateral gates in Fig. 32 and source and drain regions in Fig. 33) is due to the truncation of the Fourier series (N = 10) and does not change significantly the results of the drain current calculation (see next section, where it is shown that the minimum of potential is the most important parameter to be reproduced accurately).

$$\varphi(x,y,z) = \varphi_{top-gate(TG)}(x,y,z) + \varphi_{back-gate(BG)}(x,y,z) +$$
$$\varphi_{lateral-gates(LG)}(x,y,z) + \varphi_{source/drain(SD)}(x,y,z)$$
(99)

$$\varphi_{TG}(x,y,z) = (V_{G1} - V_{FB1})\sum_{m=1}^{+\infty}\sum_{n=1}^{+\infty} F_C(m)F_C(n)\sin(\frac{m\pi x}{W})\sin(\frac{n\pi z}{L_G}) \times$$
$$\left[\cosh(\sqrt{(\frac{m}{W})^2 + (\frac{n}{L_G})^2}\pi y) \Big/ \cosh(\sqrt{(\frac{m}{W})^2 + (\frac{n}{L_G})^2}\pi(H + t_{ov})) \right]$$
(100)

$$\varphi_{LG}(x,y,z) = (V_{G1} - V_{FB1})\sum_{m=1}^{+\infty}\sum_{n=0}^{+\infty} F_P(m)F_n(n)\sin(\frac{m\pi z}{L_G})\sin(\frac{(2n+1)\pi(H + t_{ov} - y)}{2(H + t_{ov})}) \times$$
$$\left[\frac{\sinh(\sqrt{(\frac{m}{L_G})^2 + (\frac{2n+1}{2(H+t_{ov})})^2}\pi(W - x)) + \sinh(\sqrt{(\frac{m}{L_G})^2 + (\frac{2n+1}{2(H+t_{ov})})^2}\pi x)}{\sinh(\sqrt{(\frac{m}{L_G})^2 + (\frac{2n+1}{2(H+t_{ov})})^2}\pi W)} \right]$$
(101)

$$\varphi_{SD}(x,y,z) = \sum_{m=1}^{+\infty}\sum_{n=0}^{+\infty} F_P(m)F_n(n)\sin(\frac{m\pi x}{W})\sin(\frac{(2n+1)\pi(H + t_{ov} - y)}{2(H + t_{ov})}) \times$$
$$\left[\frac{V_S \sinh(\sqrt{(\frac{m}{W})^2 + (\frac{2n+1}{2(H+t_{ov})})^2}\pi(L_G - z)) + V_D \sinh(\sqrt{(\frac{m}{W})^2 + (\frac{2n+1}{2(H+t_{ov})})^2}\pi z)}{\sinh(\sqrt{(\frac{m}{W})^2 + (\frac{2n+1}{2(H+t_{ov})})^2}\pi L_G)} \right]$$
(102)

$$\varphi_{BG}(x,y,z) = \varphi_{S3}\sum_{m=1}^{+\infty}\sum_{n=1}^{+\infty} F_P(m)F_P(n)\sin(\frac{m\pi x}{W})\sin(\frac{n\pi z}{L_G}) \times$$
$$\left[\sinh(\sqrt{(\frac{m}{W})^2 + (\frac{n}{L_G})^2}\pi(H + t_{ov} - y)) \Big/ \sinh(\sqrt{(\frac{m}{W})^2 + (\frac{n}{L_G})^2}\pi(H + t_{ov})) \right]$$
(103)

with:

$$\varphi_{S3} = (V_{G2} - V_{FB2}) \Bigg/ \left(1 + \frac{\varepsilon_{Si}}{\varepsilon_{BOX}/(t_{OX2} - t_{ov})} \sum_{m=1}^{+\infty} \sum_{n=1}^{+\infty} F_P(m) F_P(n) \sin(\frac{m\pi}{2}) \sin(\frac{n\pi}{2}) \times \left[\sqrt{(\frac{m}{W})^2 + (\frac{n}{L_G})^2} \, \pi \Bigg/ \tanh(\sqrt{(\frac{m}{W})^2 + (\frac{n}{L_G})^2} \, \pi(H + t_{ov})) \right] \right) \tag{104}$$

W being the fin width, H the fin height, L_G the channel length, t_{ov} the overetch depth into the BOX, ε_{BOX} the BOX permittivity, ε_{Si} the silicon permittivity, V_{FB1} (resp. V_{FB2}) the front-gate (resp. back-gate) flat band voltage. The series coefficients F_C, F_P, and F_N are defined in the Appendix.

Additionally, it should be noted that the potential value in ΠFETs is higher than for TGFETs (Fig. 32 and Fig. 33). This highlights the better electrostatic control from the front-gate in ΠFETs architectures. This enhanced electrostatic control of the gate on the channel limits the influence of the source and drain, and accounts for improved Short-Channel Effects (SCEs) in the TGFETs and ΠFETs architectures as compared with single-gate and double-gate SOI MOSFETs.

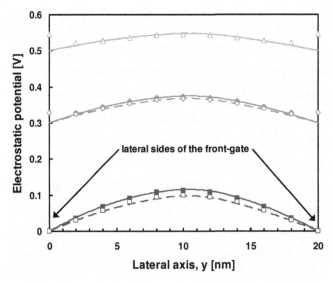

Fig. 32: Transversal cross-section at mid-channel length ($z = L_G/2$) and at the body/BOX interface ($y = t_{ov}$) of the electrostatic potential obtained with a numerical simulator (solid lines, using COMSOL Femlab [56]) and with the analytical model for TGFETs (closed symbols) and ΠFETs (open symbols). Front-gate bias $V_{G1} = 0$ V (squares), 0.3 V (diamonds) and 0.5 V (triangles). $L_G = 40$ nm, W = 20 nm, H = 20 nm, $t_{ov} = 20$ nm, $t_{OX1} = 2$ nm, $t_{OX2} = 100$ nm, $V_{DS} = 0.8$ V, $V_{G2} = 0$ V.

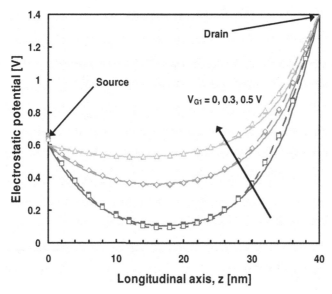

Fig. 33: Transversal cross-section at mid-channel width (x = W/2) and at the body/BOX interface (y = t_{OV}) of the electrostatic potential obtained with a numerical simulator (solid lines, using COMSOL Femlab [56]) and with the analytical model for TGFETs (closed symbols) and ΠFETs (open symbols). Front-gate bias V_{G1} = 0 V (squares), 0.3 V (diamonds) and 0.5 V (triangles). L_G = 40 nm, W = 20 nm, H = 20 nm, t_{OV} = 20 nm, t_{OX1} = 2 nm, t_{OX2} = 100 nm, V_{DS} = 0.8 V, V_{G2} = 0 V.

3.3.2. *Calculation of the subthreshold current*

Position of the minimum of potential

Once the potential profile obtained, several approximations are necessary in order to derive the subthreshold characteristics. It is often considered that the subthreshold current is flowing preferentially in the region where the electrostatic control of the front-gate is the weakest (the so-called 'most leaky path' approach). Therefore, calculating the minimum of the potential barrier and its location, the subthreshold characteristics of the transistor can be derived (a detailed description of the procedure can be found in the BSIM-CMG model for planar symmetrical DGFETs [45]). In TGFET and ΠFET devices where all the sides of the front-gate are connected, the minimum of potential is located at mid-channel (x = W/2, Fig. 31) due to obvious symmetry considerations and at the body/overetched region interface (y = t_{OV}, Fig. 31).

Concerning the position of the minimum potential value along the source drain axis (z axis), the approximate formula given in [57] has been used:

$$Z_C = \frac{L_G}{2} + \frac{L_D}{2\pi} \ln(\frac{-\varphi_{MS}}{-\varphi_{MS} + V_{DS}})$$
(105)

where:

$$L_D = \left(\frac{1}{W^2} + \frac{0.5}{H^2} \right)^{-1/2} \tag{106}$$

with L_G the channel length, W the channel width, H the channel height, φ_{MS} the gate work function, V_{DS} the drain bias, and L_D a decay length in the TGFETs.

The location of the minimum of potential along the source/drain axis obtained with Eqs. (105)-(106) has been compared with the results obtained via numerical simulations (Fig. 34). The formula is able to reproduce correctly the position of Z_C; the slight difference does not change significantly the obtained subthreshold current. It is important to note as well that, in the Π architecture, the value of the minimum potential is marginally changed, but its location is perfectly reproduced (Fig. 33). Therefore, the equations (105)-(106) derived for TGFETs can be directly used.

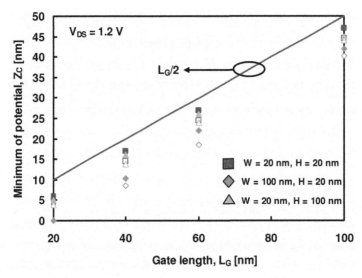

Fig. 34: Minimum of potential vs. channel length L_G for various channel width W and height H obtained with the numerical simulations (closed symbols) and the analytical formula (open symbols). The case of $Z_C = L_G /2$ corresponding to low V_{DS} values is also shown for comparison (solid line).

Subthreshold current

By assuming drift-diffusion transport [58], the current flowing from source to drain can be expressed as:

$$I_{DS}(W,H) = q n_i \mu \frac{\int_0^{V_{DS}} e^{-\varphi_F/V_t} d\varphi_F}{\int_0^L \frac{dx}{\int_0^{WH+t_{ov}} \int_{t_{ov}} e^{\varphi(x,y,z)/V_t} dy dx}} \tag{107}$$

with q the electron charge, μ the mobility, n_i the intrinsic concentration, Zc the position of the minimum of potential along the S/D axis, and $V_t = kT/q$ the thermal voltage. The role of interface traps can be safely ignored as being screened by the large capacitance of the gate dielectric ($qD_{it} << C_{ox}$).

Making the approximation that the Fermi potential is constant between source and drain and that the current is described by its value where the potential along the source/drain axis is minimal (the 'most leaky path' approach), Eq. (107) can be rewritten as:

$$I_{DS}(W,H) = \frac{qn_i\mu V_t}{L_G}(1-e^{-V_{DS}/V_t}) \int_0^W\int_0^H e^{\phi(x,y,Z_C)/V_t} dxdy \tag{108}$$

with the parameters described in (49).

The double integration along the width and height of the transistor in Eq. (108) should be simplified in order to get a tractable solution. The potential along the width direction being a parabola, the exponential of the potential is a Gaussian curve (Fig. 35). In this work, we performed a Lagrange interpolation of the *exponential* of the potential; we kept the second order to get a tractable equation. A better precision can be obtained using 3rd or 4th order approximation, but the additional terms make the current formula far less tractable. Along the vertical axis, the potential has been assumed to be equal to its value at the back-interface ($y = t_{ov}$). Therefore, the *exponential* of the electrostatic potential is correctly interpolated in the vicinity of the potential minimum ($x = W/2$, $y = t_{ov}$, $z = Z_C$, Fig. 35). These approximations enable us to calculate the leakage current in the subthreshold regime considering that it is dominated by the flow of carriers where the electrostatic control of the front-gate is the weakest.

Using the above approximations, Eq.(108) can be directly integrated leading to the subthreshold current expressed as:

$$I_{DS}(V_{G1},V_{DS}) = \frac{qn_i\mu V_t}{L_G}(1-e^{-V_{DS}/V_t})\left[WH\frac{2e^{\phi(\frac{W}{2},e_{ov},Z_C)/V_t}+e^{V_{G1}/V_t}}{3}\right] \tag{109}$$

with Zc defined in (47)-(48) and $\varphi(W/2, t_{ov}, Z_C)$ obtained from (41)-(46). Please note that $\varphi(W/2, t_{ov}, Z_C)$ is also a function of the front-gate bias V_{G1}.

The subthreshold current model (Eq. (109)) is compared with experimental data (Fig. 36) for ΩFET devices [45][46], where the gate encroachment *under* the channel is small enough ($\approx$ 5 nm, Fig. 31) to consider them as virtually ΠFETs. The obtained agreement is good (Fig. 36) for a variety of channel lengths and drain biases in the linear regime and in saturation. The formula is not compact in the sense that it uses a series development (for the value of φ(W/2,t_{OV}, Z_C)) but a truncation of the series to the 5th term does not lead to significant deviations compared to the complete series.

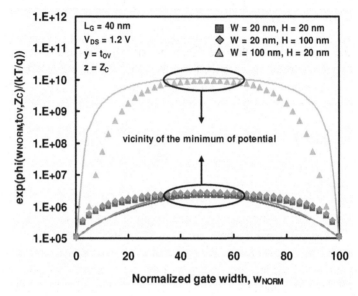

Fig. 35: Value of the exponential of the potential in Eq. (50) vs. normalized position along the channel width W_{NORM} at the channel/BOX interface (y = t_{OV}) and at the minimum of potential (z= Z_C) obtained from numerical simulation (solid lines) and approximation made with the Lagrange interpolation of the analytical model (symbols). Comparison made for W = 20 nm / H = 20 nm (squares), W = 20 nm / H = 100 nm (diamonds), and W = 100 nm / H = 20 nm (triangles). V_{DS} = 1.2 V, t_{OX1} = 2 nm, t_{OX2} = 100 nm, t_{OV} = 30 nm.

Subthreshold slope, Roll-off, DIBL

The analytical subthreshold slope can be extracted using (51) and considering two gate biases in the subthreshold regime (an approximate simpler analytical expression can be derived as well, see section 3.3.3.). A good agreement is found with the experimental measurements (Fig. 37). It is shown that the scaling of the channel width W together with the channel length L_G allows maintaining the SCEs under control. Additionally, for a given channel length L_G, reducing the channel width W leads to a better electrostatic control of the front-gate on the channel, and therefore to an improved subthreshold slope.

The threshold voltage in the linear regime and in saturation has been extracted using the constant current method (Fig. 38); a convincing agreement is found. It can be seen that the V_{TH} decreases with the channel length when the devices are operated in the saturation regime. The Roll-off effect is accurately reproduced.

Similarly, the variation of the threshold voltage between low and high values of the drain voltage yields straightforwardly the DIBL effect. Similar trends as for the subthreshold slope are observed, and a correct agreement is obtained (Fig. 39).

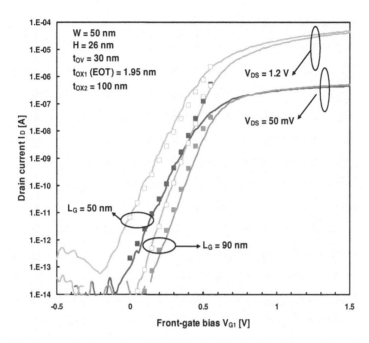

Fig. 36: Analytical (symbols) and experimental (solid lines) subthreshold drain current I_{DS} vs. front-gate bias V_{G1} for channel lengths L_G of 90 nm (squares) and 50 nm (diamonds). Channel width W = 50 nm, channel height H = 26 nm, t_{OX1}=1.95 nm, t_{OX2} = 100 nm.

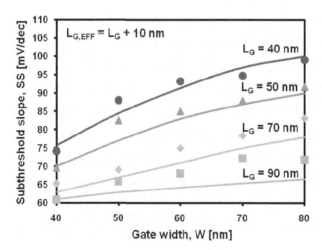

Fig. 37: Subthreshold slope vs. channel width W for channel lengths L_G of 90, 70, 50, and 40 nm. Comparison between experimental extractions (symbols) and our analytical model (solid lines). t_{OX1}=1.95 nm, t_{OX2} = 100 nm, H = 26 nm. V_{DS} = 5 mV.

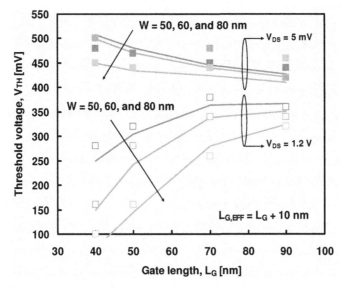

Fig. 38: Threshold voltage V_{TH} extracted with the constant current method (at 0.1 µA, symbols) and with the analytical model (solid lines) vs. channel length L_G. The channel width W varies from 50 to 80 nm, and V_{DS} from 5 mV (closed symbols) to 1.2 V (open symbols t_{OX1}=1.95 nm, t_{OX2} = 100 nm, H = 26 nm.

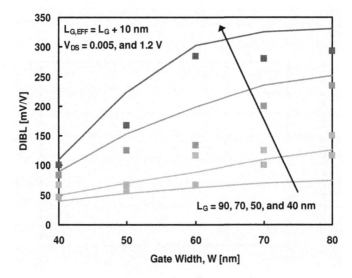

Fig. 39: DIBL vs. channel width W for channel lengths L_G of 90, 70, 50, and 40 nm. Comparison between experimental extractions (symbols) and analytical model (solid lines). t_{OX1}=1.95 nm, t_{OX2} = 100 nm, H = 26 nm.

3.3.3. *Device scalability*

The ΠFET structure is very informative since it intrinsically contains all the other multi-gate transistors by simply changing the device parameters (see Tab. 2). Therefore, the analytical formulas derived for ΠFETs can be used as a core model to depict the subthreshold slope SS vs. channel length L_G for a wide range of devices (Fig. 40, with the notable exception of ΩFET transistors). Taking simulation data from [43] and our own experimental measurements for planar FDSOI wide devices, an excellent agreement with the model is found. As highlighted in [43][45], increasing the 'number of gates' allows a better scalability of the MuGFET SOI devices. ΠFETs, acting like a device with a number of gates between 3 and 4 (depending of the overetch depth and channel width), offers increased front-gate control, closer to the one in a 4-gate Gate-All-Around (GAA FETs) device. The advantage of ΠFETs comes from a very pragmatic process flow as opposed to 4-gate Gate-All-Around devices. Compared to a traditional TGFET process flow, the channel etch step duration only needs to be extended.

Table 2. Variations of the core structure

Structure	Features
ΠFET (core structure)	$t_{OV} \neq 0$
TGFET	$t_{OV} \approx 0$
Planar FDSOI	$t_{OV} \approx 0$, W>>H
DGFET/FinFET	$t_{OV} \approx 0$, W<<H
Gate All Around	$t_{OV} \approx 0$, $\varphi_{S3} = V_{G1} - V_{FB1}$

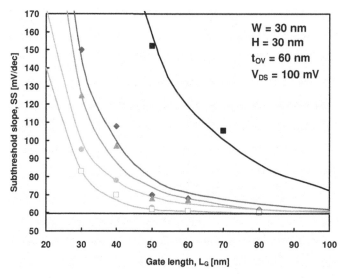

Fig. 40: Subthreshold slope SS vs. channel length L_G obtained with numerical simulations (symbols) and the analytical model (solid lines) for planar FDSOI FETs (squares), DGFET (diamonds), TGFET (triangles), ΠFET (circles), and GAA (open squares) transistors. All numerical simulation results are obtained from [20], except for planar FDSOI FETs (H = 26 nm, W = 10 μm), where our experimental measurements are used.

Additionally, the silicon channel width/height necessary to keep the SCEs under control (using a criterion of 75 mV/dec subthreshold swing) is calculated as a function of the channel length for MuGFETs (Fig. 41). The better scalability achieved by increasing the number of gates is highlighted once again, as the necessary dimensions of the width/height of the channel compared to the channel length are relaxed; for example, at L_G = 60 nm the ratios W/L_G or H/L_G change from 0.25 (in planar FDSOI devices) to 1.2 (in GAA). It is noteworthy that the scaling rules presented in [45] based on the 'natural length' concepts and derived from a solution of the Poisson's equation assuming a simplified potential shape are also represented (coloured areas, Fig. 41). They are remarkably coherent with the formula developed in this work.

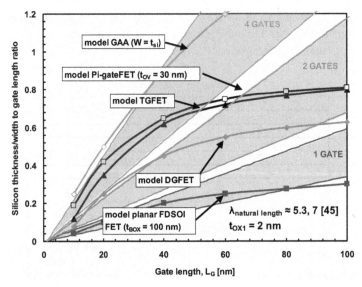

Fig. 41: Silicon width/thickness to channel length ratio vs. channel length L_G for FDSOI, DGFET, TGFET, ΠFET, and GAA transistors (solid lines) obtained for a subthreshold slope criterion of 75 mV/dec. Coloured areas show the results obtained with the 'natural length' approaches [22]. t_{OX1} = 2 nm, t_{OX2} = 100 nm, V_{DS} = 100 mV.

Subthreshold slope pseudo-compact formula

The goal of this section is to derive a simplified analytical expression for the subthreshold slope, without using the analytical formula of the subthreshold current (51) in order to provide fast usable scalability guidelines. In Eq. (109), it was found that the second term in the brackets (V_{G1} dependent) was less important that the first term (potential dependent) for not too short channels (i.e., respecting the scaling rules). Therefore the subthreshold current Eq. (109) could be approximately written as:

$$I_{DS}(V_{G1}, V_{DS}) \approx \frac{q n_i \mu V_t}{L_G} (1 - e^{-V_{DS}/V_t}) WH \frac{2^{\phi(\frac{W}{2}, t_{ov}, z_C)/V_t}}{3} \tag{110}$$

Therefore, the subthreshold slope can be directly expressed as:

$$SS(L_G, W, H) = (\frac{\partial \ln(I_D)}{\partial V_G})^{-1} = V_T \ln(10) \times$$

$$\dfrac{1}{\left[\displaystyle\sum_{m=1}^{N} \sum_{n=1}^{N} \dfrac{F_C(m) F_C(n) \sin(\dfrac{m\pi}{2}) \sin(\dfrac{n\pi Z_C}{2})}{\cosh(\sqrt{(\dfrac{m}{W})^2 + (\dfrac{n}{L_G})^2} \, \pi H)} + \displaystyle\sum_{m=1}^{N} \sum_{n=0}^{N} F_P(m) F_N(n) \sin(\dfrac{m\pi Z_C}{L})(-1)^n \dfrac{2\sinh(\sqrt{(\dfrac{m}{L_G})^2 + (\dfrac{2n+1}{2H})^2} \, \pi \dfrac{W}{2})}{\sinh(\sqrt{(\dfrac{m}{L_G})^2 + (\dfrac{2n+1}{2H})^2} \, \pi W)} \right]} \qquad (111)$$

with Z_C defined in (105)-(106) and F_C, F_P, F_N in appendix A.

The approximation performed to derive Eq.(111) degrades the accuracy (error between the numerical simulation and Eq.(111) ranging from 0 to 20%), but remains acceptable (Fig. 42). Equation Eq.(111) can also be extended to the case of FinFETs/DGFETs and planar FDSOI devices (Fig. 42), with a very correct precision. The formula does not use any fitting parameters (unlike in references [57][58], where the position of the most leaky path is adjusted and said to move with the drain bias). The formula is even *pseudo-compact*, considering that a truncation at N=5 gives no significant difference with the calculations made for N $\rightarrow$ +∞ (Fig. 43).

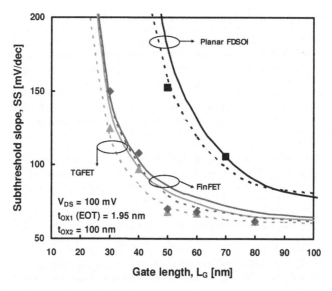

Fig. 42: Comparison between the subthreshold swing obtained with the analytical solution (dashed lines) and the expression proposed in (53) (solid lines). The symbols represent the numerical simulation results obtained for TGFET (W= 30 nm, H = 30 nm, triangles) and FinFET (W = 30 nm, H = 100 nm, diamonds) and the experimental extraction for Planar FDSOI devices (squares, W = 10 μm, H = 26 nm).

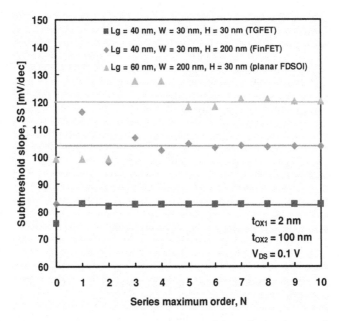

Fig. 43: Subthreshold slope given by Eq.(53) as a function of the series maximum order N for TGFET (squares), FinFETs (diamonds), and planar FDSOI MOSFET (triangles). $t_{OX1} = 2$ nm, $t_{OX2} = 100$ nm, $V_{DS} = 0.1$ V.

3.4. *Conclusions*

This part dealt with the modeling of Triple-gate structures and their variants such as ΠFETs and ΩFETs.

First, analytical models for the long-channel threshold voltage of ΩFETs, ΠFETs, Triple-gateFETs, and planar FDSOI transistors were presented. The models are valid for the whole range of back-gate biases (back-interface accumulation, depletion, and inversion regimes), and for NMOS/PMOS. The validation with both numerical simulations and measurements yields a very good agreement. Using the model derived for long channels ΠFET, TGFETs and planar FDSOI transistors, it is demonstrated that experimental fully depleted devices can operate in the 'back-interface inversion' regime even at V_{G2} grounded. As a result, two threshold voltages appear in the transistors, with an experimental difference of threshold voltages up to several hundreds mV for planar FDSOI devices. As a consequence, the back-interface channel is activated, leading to off-state current increase and subthreshold slope degradation. This effect can be alleviated by keeping a thick BOX when scaling down the structures, or by using vertical Multiple-GateFETs.

Concerning short-channels, an analytical model for the subthreshold performance of ΠFET was developed. It is shown that the effect of the gate penetration in the BOX is of great importance since it improves significantly the subthreshold performance of the device. Based on the solution of the 3D potential profile obtained with the Laplace's equation, the value and position of the minimum of potential is calculated. Then, the subthreshold current is obtained analytically without the need of any fitting parameters, after simplification of the integral of the potential along the section of the 'most leaky path'. The subthreshold slope, threshold voltage, and DIBL are then extracted from experimental devices (featuring a channel length L_G between 90 and 40 nm, a channel width W between 80 and 40 nm, and a channel height H of 26 nm) and favorably compared to the model.

Additionally, it is demonstrated that without the use of any fitting parameter, the model can be extended to nearly all the MuGFETs devices (GAA/ΠFETs/TGFETs/FinFETs/symDGFETs/FDSOI planar devices), and therefore can be used as a core model for the scaling and calibration of a wide range of MuGFETs. A 3D *pseudo-compact* subthreshold slope formula (approximation leading to an error ranging between 0 and 20 %) is also proposed for ΠFETs, TGFETs, FinFETs, symDGFETs or FDSOI planar devices, in order to provide a direct and easy formula for anticipating the scaling of MuGFET transistors.

Appendix A. Fourier Series Coefficients

- Constant potential boundary condition :

$$F_C(n) = \frac{2(1-\cos(n\pi))}{n\pi}$$

- Parabolic potential boundary condition

$$F_P(n) = \frac{2(1-\cos(n\pi)) - n\pi\sin(n\pi)}{(\frac{n\pi}{2})^3}$$

- Neumann boundary condition :

$$F_N(n) = \frac{4}{(2n+1)\pi}(1-\cos(\frac{(2n+1)\pi}{2}))$$

Acknowledgments

This work was supported by the European Union IAPP FP7 'Marie Curie' COMON (COmpact MOdeling Network) project (ref. pro. 218255). Support from NANOSIL (contract 216171) and EUROSOI+ (contract 216373) contracts of the European Commission, the Spanish Ministry of Science under TEC2008-06758-C02-02/TEC, PGIR/15 grant from URV and ICREA Academia Prize are also greatly acknowledged.

The authors would also like to thank Dr. Mingchun Tang (formerly at Université de Strasbourg – UdS, France) for fruitful discussions and Dr. J. Buckley (CEA-LETI Grenoble, France) for correcting mistakes.

References

1. F. Lime, B. Iñiguez, and O. Moldovan, A quasi-two-dimensional compact drain-current model for undoped symmetric double-gate MOSFETs including short-channel effects, *IEEE Transactions on Electron Devices* **55**(6), 1441-1448 (2008).

2. J.-M. Sallese, F. Krummenacher, F. Prégaldiny, C. Lallement, A. Roy, and C. Enz, A design oriented charge-based current model for symmetric DG MOSFET and its correlation with the EKV formalism, *Solid-State Electronics* **49**, 485-489 (2005).

3. B. Iñiguez, D. Jiménez, J. Roig, H.A. Hamid, L.F. Marsal, and J. Pallarès, Explicit continuous model for long-channel undoped surrounding gate MOSFETs, *IEEE Transactions on Electron Devices* **52**(8), 1868-1873 (2005).

4. F. Lime, R. Ritzenthaler, M. Ricoma, F. Martinez, F. Pascal, E. Miranda, O. Faynot and B. Iñiguez, A Physical Compact DC Drain Current Model for Long-Channel Undoped Ultrathin Body (UTB) SOI and Asymmetric Double-Gate (DG) MOSFETs with Independent Gate Operation, *Solid-State Electronics* **57**, 61-66 (2011)

5. O. Moldovan, F.A. Chaves, D. Jiménez, and B. Iñiguez, Compact charge and capacitance modeling of undoped ultra-thin body (UTB) SOI MOSFETs, *Solid State Electronics* **52**, 1867-1871 (2008).

6. A.S. Roy, J.M Sallese, and C.C. Enz, A closed-form charge-based expression for drain current in symmetric and asymmetric double gate MOSFET, *Solid State Electronics* **50**, 687-693 (2006).

7. M. Reyboz, P. Martin, T. Poiroux, O. Rozeau, Continuous model for independent double gate MOSFET, *Solid-State Electronics* **53**(5), 504-513 (2009).

8. F. Liu, J.He, Y.Fu, J.Hu, W.Bian, Y. Song, and X. Zhang, Generic Carrier-Based Core Model for Undoped Four-Terminal Double-Gate MOSFETs Valid for Symmetric, Asymmetric, and Independent-Gate-Operation Modes, *IEEE Transactions on Electron Devices* **55**(3), 816-826 (2008).

9. A. Ortiz-Conde and F. J. García Sánchez, Unification of asymmetric DG, symmetric DG and bulk undoped-body MOSFET drain current, Solid-State Electronics **50**, 1796-1800 (2006).

10. H.-K. Lim and J.G. Fossum, Threshold voltage of thin-film silicon-on-insulator (SOI) MOSFET's, *IEEE Transactions on Electron Devices* **ED-30**(10), 1244-1251 (1983).

11. H. Lu and Y. Taur, An analytical potential model for symmetric and asymmetric DG MOSFETs, *IEEE Transactions on Electron Devices* **53**(5), May 2006.

12. A. Ortiz-Conde, F. J. García-Sánchez, J. Muci, and S. Malobabic, A general analytical solution to the one-dimensional undoped oxide-silicon-oxide system, in *Proc. 6th Int. Caribbean Conf. Device, Circuits Syst.*, 177–182 (2006).

13. A. Ortiz-Conde, F. J. García-Sánchez, S. Malobabic, J. Muci, and R. Salazar, Drain-current and transconductance model for the undoped body asymmetric double-gate MOSFET, in *Proc. 8th Int. Conf. Solid-State Integr.-Circuit Technol.*, 1239-1242 (2006).

14. G. Dessai, W. Wu and G. Gildenblat, Compact charge model for independent-gate asymmetric DGFET, *IEEE Transactions on Electron Devices* **57**(9), 2106-2115 (2010).

15. V. Barral et al., Strained FDSOI CMOS technology scalability down to 2.5nm film thickness and 18nm gate length with a TiN/HfO2 gate stack, in *Proceedings of International Electron Device Meetings IEDM 2007*, (2007).

16. B. Iniguez and A. Garcia, An Improved C -Continuous Small-Geometry MOSFET Modeling for Analog Applications, *Analog Integrated Circuits and Signal Processing* **13**, 241–259 (1997).

17. Y.A. El-Mansy and A.R. Boothroyd, A simple two-dimensional model for IGFET operation in the saturation region, *IEEE Trans. Electron Devices*, **24**(3), 254-261 (1977).

18. C.C. McAndrews, B.K. Bhattacharyya and O. Wing, A single-piece C-continuous MOSFET model including subthreshold conduction, *IEEE Electron Device Letters* **12**, 565-567 (1991).

19. K.K. Young, Short-channel effects in fully depleted SOI MOSFETs, *IEEE Trans. Electron Devices*, **36**(2), 399-402 (1989).

20. R.H. Yan, A. Ourmazd, K.F. Lee, Scaling the Si MOSFET: from bulk to SOI to bulk, *IEEE Trans. Electron Devices*, **39**(7), 1704-1710 (1992)

21. K. Suzuki, T. Tanaka, Y. Tosaka, H. Horie, Y. Arimoto, Scaling theory for double-gate SOI MOSFET's, *IEEE Trans. Electron Devices*, **40**(12), 2326-2329 (1993).

22. H.C. Chow, W.S. Feng, An Improved Analytical Model for Short-Channel MOSFET's, *IEEE Trans. Electron Devices*, **39**(11), 2626-2629 (1992).

23. D. Jiménez, B. Iñiguez, J. Suñé, L. F. Marsal, J. Pallarès, J. Roig and D. Flores, Comment on "New current-voltage model for surrounding-gate metal oxide semiconductor field-effect transistors", [Jpn. J. Appl. Phys. 44 (2005) 6446], *Japanese Journal of Applied Physics Part 1-Regular Papers Brief Communications & Review Papers* **45**, 6057, (2006).

24. International Technology Roadmap for Semiconductors (ITRS), 2010 edition, available: http://www.itrs.net/Links/2010ITRS/2010Update/ToPost/2010_Update_Overview.pdf

25. H. Lee et al., "Sub-5nm All-Around Gate FinFET for Ultimate Scaling," Digest of Technical Papers, *2006 Symposium on VLSI Technology*, pp. 58-59, 2006.

26. F.-L. Yang et al., "5nm-Gate Nanowire FinFET," Digest of Technical Papers, *2004 Symposium on VLSI Technology*, pp. 196-197, 2004.

27. C. W. Mueller, P. H. Robinson, "Grown-film silicon transistors on sapphire", *proceedings of the IEEE*, vol. 52, no. 12, pp. 1497-1490, 1964.

28. F. Assaderaghi, D. Sinitsky, S. Parke, J. Bokor, P. K. Ko, C. Hu, "A Dynamic Threshold Voltage MOSFET (DTMOS) for Ultra-Low Voltage Operation", *IEDM'94 Technical Digest*, pp. 809-812, 1994.

29. T. Douseki, N. Shibata, J. Yamada, "SIMOX ROM Macro with Low-Vth Memory Cells", *IEEE Int. SOI conf. 2001*, pp. 143-144, 2001.

30. J.-P. Colinge, "Thin-film, accumulation-mode p-channel SOI MOSFETs", *Electronic letters*, Vol. 24, no. 5, 3 March 1988, pp. 257-258, 1988.

31. S. Cristoloveanu and S.S. Li, "electrical characterization of silicon-on-insulator materials and devices", Kluwer Academic Publishers, ISBN 978-0792395485, 1995.

32. J.-P. Colinge, M.H. Gao, A. Romano-Rodriguez, H. Maes, C. Claeys, "Silicon-on-insulator "Gate-All-Around Device" ", *IEDM'90 Technical Digest*, pp. 595-598, 1990.

33. T. Ernst et al., "Novel 3D integration process for highly scalable Nano-Beam stacked-channels GAA (NBG) CMOSFETs with HfO2/TiN gate stack", *IEDM'06 Technical Digest*, 2006.

34. D. Hisamoto, T. Kaga, Y. Kawamoto, E. Takeda, "A fully depleted lean-channel transistor (DELTA)-a novel vertical ultra thin SOI MOSFET", *IEDM'94 Technical Digest*, pp. 833-836, 1989.

35. Y. X. Liu, M. Masahara, K. Ishii, T. Tsutsumi, T. Sekigawa, H. Takashima, H. Yamauchi, E. Suzuki, "Flexible threshold voltage FinFETs with independent double gates and an ideal rectangular cross-section Si-Fin channel", *IEDM'03 Technical Digest*, pp. 986-988, 2003.

36. S. Miyano, M. Hirose, F. Masuoka, "Numerical analysis of a cylindrical thin-pillar transistor (CYNTHIA)", *Electron Devices, IEEE Transactions on*, Vol. 39, no. 8, pp. 1876-1881, 1992.

37. D. Jiménez, B. Iñiguez, J. Suñé, L. F. Marsal, J. Pallarès, J. Roig, D. Flores, "Continuous Analytical I-V Model for Surronding-Gate MOSFETs", *IEEE Electron Device Letters*, vol. 25, no. 8, pp. 571-573, 2004.

38. J.-P. Colinge, "Multiple-gate SOI MOSFETs", *Solid State Electronics*, vol. 48, no. 6, pp. 897-905, 2004.

39. D. Hisamoto et al., "FinFET – A Self-Aligned Double-Gate MOSFET Scalable to 20 nm," *Electron Devices, IEEE Transactions on*, vol. 47, no. 12, pp. 2320-2325, Dec. 2000.

40. B. Doyle et al., "Tri-Gate Fully Depleted CMOS Transistors: Fabrication, Design and Layout," *Digest of Technical Papers, 2003 Symposium on VLSI Technology*, pp. 132-133, 2003.

41. J. G. Fossum, J.-W. Wang, and V. P. Trivedi, "Suppression of corner effects in Triple-gate MOSFETs", *IEEE Electron Device Letters*, vol. 24, no. 12, pp. 745-747, Dec. 2003.

42. R. Ritzenthaler, O. Faynot, T. Poiroux, C. Jahan, and S. Cristoloveanu, "Position-Dependent Threshold in FinFETs and Triple-gate FETs," proceedings of *EUROSOI Network 2nd Workshop on Silicon-On-Insulator Technology Devices and Circuits*, 2006.

43. J.-T. Park, J.-P. Colinge, and C.H. Diaz, "Pi-Gate SOI MOSFET," *IEEE Electron Device Letters*, vol. 22, no. 8, pp. 405-406, Aug. 2001.

44. F.-L. Yang et al., "25 nm CMOS Omega FETs", *IEDM'02 Technical Digest*, pp. 255-258, 2002.

45. J.-P. Colinge et al., "FinFETs and Other Multi-Gate Transistors," Springer, ISBN 978-0-387-71751-7, 2007.

46. C. Jahan et al., "10nm ΩFETs transistors with TiN metal gate and HfO2," *Digest of Technical Papers, 2005 Symposium on VLSI Technology*, pp. 112-113, 2005.

47. K. Akarvardar et al., "A Two-Dimensional Model for Interface Coupling in Triple-Gate Transistors," *Electron Devices, IEEE Transactions on*, vol. 54, no. 4, pp. 767-775, Apr. 2007.

48. J.-P. Colinge, J. C. Alderman, W. Xiong, and C. Rinn Cleavelin, "Quantum-Mechanical Effects in Trigate SOI MOSFETs," *Electron Devices, IEEE Transactions on*, vol. 53, no. 5, pp. 1131-1136, May 2006.

49. K.-T. Tang, "Mathematical method for engineers and scientists 3," Springer, ISBN 978-3540446958, 2007.

50. H.K. Lim, and J.G. Fossum, "Threshold voltage of thin-film silicon-on-insulator (SOI) MOSFETs," *Electron Devices, IEEE Transactions on*, vol. 30, p1244, 1983.

51. R. Ritzenthaler et al., "Lateral Coupling and Immunity to Substrate Effect in ΩFET Devices," *Solid-State Electronics*, vol. 50, no. 4, pp. 558-565, Apr. 2006.

52. ATLAS Device Simulation Framework, User's Manual, Silvaco Int., Santa Clara, CA, 2002.

53. F. Lime, R. Ritzenthaler, M. Ricoma, F. Martinez, F. Pascal, E. Miranda, O. Faynot and B. Iñiguez, "A Simplified Physical DC Model for Undoped UTB SOI and Asymmetric DGMOSFETs with Independent Gate Operation," *Solid State Electronics*, 57 (2011), pp. 61-66.

54. T. Ernst et al., "A Model of Fringing Fields in Short-Channel Planar and Triple-Gate SOI MOSFETs," *Electron Devices, IEEE Transactions on*, vol. 54, no. 6, pp. 1366-1375, 2007.

55. F. J. Garcia Ruiz, A. Godoy, F. Gamiz, C. Sampedro, and L. Donetti, "A Comprehensive Study of the Corner Effects in Pi-Gate MOSFETs Including Quantum Effects", *Electron Devices, IEEE Transactions on*, vol. 54, no. 12, pp. 3369-3377, Dec. 2007.

56. COMSOL Multiphysics [online]. Available: http://www.comsol.com.

57. G. Pei et al., "FinFET Design Considerations Based on 3D Simulation and Analytical Modeling," *Electron Devices, IEEE Transactions on*, vol. 49, no. 8, pp. 1411-1419, Aug. 2002.

58. H. Abd El Hamid et al., "A 3D Analytical Physically Based Model for the Subthreshold Swing in Undoped Trigate FinFETs," *Electron Devices, IEEE Transactions on*, vol. 54, no. 9, pp. 2487-2496, Sept. 2007.

AUTHOR INDEX

Printed in the United States
By Bookmasters